Human Anatomy: A Very Short Introduction

VERY SHORT INTRODUCTIONS are for anyone wanting a stimulating and accessible way into a new subject. They are written by experts, and have been translated into more than 40 different languages.

The series began in 1995, and now covers a wide variety of topics in every discipline. The VSI library now contains over 350 volumes—a Very Short Introduction to everything from Psychology and Philosophy of Science to American History and Relativity—and continues to grow in every subject area.

Very Short Introductions available now:

ACCOUNTING Christopher Nobes
ADVERTISING Winston Fletcher
AFRICAN AMERICAN RELIGION Eddie S. Glaude Jr.
AFRICAN HISTORY John Parker and Richard Rathbone
AFRICAN RELIGIONS Jacob K. Olupona
AGNOSTICISM Robin Le Poidevin
ALEXANDER THE GREAT Hugh Bowden
AMERICAN HISTORY Paul S. Boyer
AMERICAN IMMIGRATION David A. Gerber
AMERICAN LEGAL HISTORY G. Edward White
AMERICAN POLITICAL HISTORY Donald Critchlow
AMERICAN POLITICAL PARTIES AND ELECTIONS L. Sandy Maisel
AMERICAN POLITICS Richard M. Valelly
THE AMERICAN PRESIDENCY Charles O. Jones
AMERICAN SLAVERY Heather Andrea Williams
THE AMERICAN WEST Stephen Aron
AMERICAN WOMEN'S HISTORY Susan Ware
ANAESTHESIA Aidan O'Donnell
ANARCHISM Colin Ward
ANCIENT EGYPT Ian Shaw
ANCIENT EGYPTIAN ART AND ARCHITECTURE Christina Riggs
ANCIENT GREECE Paul Cartledge
THE ANCIENT NEAR EAST Amanda H. Podany
ANCIENT PHILOSOPHY Julia Annas
ANCIENT WARFARE Harry Sidebottom
ANGELS David Albert Jones
ANGLICANISM Mark Chapman
THE ANGLO-SAXON AGE John Blair
THE ANIMAL KINGDOM Peter Holland
ANIMAL RIGHTS David DeGrazia
THE ANTARCTIC Klaus Dodds
ANTISEMITISM Steven Beller
ANXIETY Daniel Freeman and Jason Freeman
THE APOCRYPHAL GOSPELS Paul Foster
ARCHAEOLOGY Paul Bahn
ARCHITECTURE Andrew Ballantyne
ARISTOCRACY William Doyle
ARISTOTLE Jonathan Barnes
ART HISTORY Dana Arnold
ART THEORY Cynthia Freeland
ASTROBIOLOGY David C. Catling
ATHEISM Julian Baggini
AUGUSTINE Henry Chadwick
AUSTRALIA Kenneth Morgan
AUTISM Uta Frith
THE AVANT GARDE David Cottington
THE AZTECS Davíd Carrasco
BACTERIA Sebastian G. B. Amyes
BARTHES Jonathan Culler
THE BEATS David Sterritt

BEAUTY Roger Scruton
BESTSELLERS John Sutherland
THE BIBLE John Riches
BIBLICAL ARCHAEOLOGY Eric H. Cline
BIOGRAPHY Hermione Lee
THE BLUES Elijah Wald
THE BOOK OF MORMON Terryl Givens
BORDERS Alexander C. Diener and Joshua Hagen
THE BRAIN Michael O'Shea
THE BRITISH CONSTITUTION Martin Loughlin
THE BRITISH EMPIRE Ashley Jackson
BRITISH POLITICS Anthony Wright
BUDDHA Michael Carrithers
BUDDHISM Damien Keown
BUDDHIST ETHICS Damien Keown
CANCER Nicholas James
CAPITALISM James Fulcher
CATHOLICISM Gerald O'Collins
CAUSATION Stephen Mumford and Rani Lill Anjum
THE CELL Terence Allen and Graham Cowling
THE CELTS Barry Cunliffe
CHAOS Leonard Smith
CHEMISTRY Peter Atkins
CHILDREN'S LITERATURE Kimberley Reynolds
CHILD PSYCHOLOGY Usha Goswami
CHINESE LITERATURE Sabina Knight
CHOICE THEORY Michael Allingham
CHRISTIAN ART Beth Williamson
CHRISTIAN ETHICS D. Stephen Long
CHRISTIANITY Linda Woodhead
CITIZENSHIP Richard Bellamy
CIVIL ENGINEERING David Muir Wood
CLASSICAL LITERATURE William Allan
CLASSICAL MYTHOLOGY Helen Morales
CLASSICS Mary Beard and John Henderson
CLAUSEWITZ Michael Howard
CLIMATE Mark Maslin
THE COLD WAR Robert McMahon
COLONIAL AMERICA Alan Taylor
COLONIAL LATIN AMERICAN LITERATURE Rolena Adorno
COMEDY Matthew Bevis
COMMUNISM Leslie Holmes
COMPLEXITY John H. Holland
THE COMPUTER Darrel Ince
CONFUCIANISM Daniel K. Gardner
THE CONQUISTADORS Matthew Restall and Felipe Fernández-Armesto
CONSCIENCE Paul Strohm
CONSCIOUSNESS Susan Blackmore
CONTEMPORARY ART Julian Stallabrass
CONTEMPORARY FICTION Robert Eaglestone
CONTINENTAL PHILOSOPHY Simon Critchley
CORAL REEFS Charles Sheppard
CORPORATE SOCIAL RESPONSIBILITY Jeremy Moon
COSMOLOGY Peter Coles
CRITICAL THEORY Stephen Eric Bronner
THE CRUSADES Christopher Tyerman
CRYPTOGRAPHY Fred Piper and Sean Murphy
THE CULTURAL REVOLUTION Richard Curt Kraus
DADA AND SURREALISM David Hopkins
DARWIN Jonathan Howard
THE DEAD SEA SCROLLS Timothy Lim
DEMOCRACY Bernard Crick
DERRIDA Simon Glendinning
DESCARTES Tom Sorell
DESERTS Nick Middleton
DESIGN John Heskett
DEVELOPMENTAL BIOLOGY Lewis Wolpert
THE DEVIL Darren Oldridge
DIASPORA Kevin Kenny
DICTIONARIES Lynda Mugglestone
DINOSAURS David Norman
DIPLOMACY Joseph M. Siracusa
DOCUMENTARY FILM Patricia Aufderheide
DREAMING J. Allan Hobson
DRUGS Leslie Iversen
DRUIDS Barry Cunliffe
EARLY MUSIC Thomas Forrest Kelly
THE EARTH Martin Redfern
ECONOMICS Partha Dasgupta
EDUCATION Gary Thomas

EGYPTIAN MYTH Geraldine Pinch
EIGHTEENTH-CENTURY BRITAIN Paul Langford
THE ELEMENTS Philip Ball
EMOTION Dylan Evans
EMPIRE Stephen Howe
ENGELS Terrell Carver
ENGINEERING David Blockley
ENGLISH LITERATURE Jonathan Bate
ENTREPRENEURSHIP Paul Westhead and Mike Wright
ENVIRONMENTAL ECONOMICS Stephen Smith
EPIDEMIOLOGY Rodolfo Saracci
ETHICS Simon Blackburn
ETHNOMUSICOLOGY Timothy Rice
THE ETRUSCANS Christopher Smith
THE EUROPEAN UNION John Pinder and Simon Usherwood
EVOLUTION Brian and Deborah Charlesworth
EXISTENTIALISM Thomas Flynn
EXPLORATION Stewart A. Weaver
THE EYE Michael Land
FAMILY LAW Jonathan Herring
FASCISM Kevin Passmore
FASHION Rebecca Arnold
FEMINISM Margaret Walters
FILM Michael Wood
FILM MUSIC Kathryn Kalinak
THE FIRST WORLD WAR Michael Howard
FOLK MUSIC Mark Slobin
FOOD John Krebs
FORENSIC PSYCHOLOGY David Canter
FORENSIC SCIENCE Jim Fraser
FOSSILS Keith Thomson
FOUCAULT Gary Gutting
FRACTALS Kenneth Falconer
FREE SPEECH Nigel Warburton
FREE WILL Thomas Pink
FRENCH LITERATURE John D. Lyons
THE FRENCH REVOLUTION William Doyle
FREUD Anthony Storr
FUNDAMENTALISM Malise Ruthven
GALAXIES John Gribbin
GALILEO Stillman Drake
GAME THEORY Ken Binmore
GANDHI Bhikhu Parekh
GENES Jonathan Slack
GENIUS Andrew Robinson
GEOGRAPHY John Matthews and David Herbert
GEOPOLITICS Klaus Dodds
GERMAN LITERATURE Nicholas Boyle
GERMAN PHILOSOPHY Andrew Bowie
GLOBAL CATASTROPHES Bill McGuire
GLOBAL ECONOMIC HISTORY Robert C. Allen
GLOBALIZATION Manfred Steger
GOD John Bowker
THE GOTHIC Nick Groom
GOVERNANCE Mark Bevir
THE GREAT DEPRESSION AND THE NEW DEAL Eric Rauchway
HABERMAS James Gordon Finlayson
HAPPINESS Daniel M. Haybron
HEGEL Peter Singer
HEIDEGGER Michael Inwood
HERODOTUS Jennifer T. Roberts
HIEROGLYPHS Penelope Wilson
HINDUISM Kim Knott
HISTORY John H. Arnold
THE HISTORY OF ASTRONOMY Michael Hoskin
THE HISTORY OF LIFE Michael Benton
THE HISTORY OF MATHEMATICS Jacqueline Stedall
THE HISTORY OF MEDICINE William Bynum
THE HISTORY OF TIME Leofranc Holford-Strevens
HIV/AIDS Alan Whiteside
HOBBES Richard Tuck
HORMONES Martin Luck
HUMAN ANATOMY Leslie Klenerman
HUMAN EVOLUTION Bernard Wood
HUMAN RIGHTS Andrew Clapham
HUMANISM Stephen Law
HUME A. J. Ayer
HUMOUR Noël Carroll
THE ICE AGE Jamie Woodward
IDEOLOGY Michael Freeden
INDIAN PHILOSOPHY Sue Hamilton
INFORMATION Luciano Floridi
INNOVATION Mark Dodgson and David Gann

INTELLIGENCE Ian J. Deary
INTERNATIONAL MIGRATION
Khalid Koser
INTERNATIONAL RELATIONS
Paul Wilkinson
INTERNATIONAL SECURITY
Christopher S. Browning
IRAN Ali M. Ansari
ISLAM Malise Ruthven
ISLAMIC HISTORY Adam Silverstein
ITALIAN LITERATURE
Peter Hainsworth and David Robey
JESUS Richard Bauckham
JOURNALISM Ian Hargreaves
JUDAISM Norman Solomon
JUNG Anthony Stevens
KABBALAH Joseph Dan
KAFKA Ritchie Robertson
KANT Roger Scruton
KEYNES Robert Skidelsky
KIERKEGAARD Patrick Gardiner
KNOWLEDGE Jennifer Nagel
THE KORAN Michael Cook
LANDSCAPE ARCHITECTURE
Ian H. Thompson
LANDSCAPES AND
GEOMORPHOLOGY
Andrew Goudie and Heather Viles
LANGUAGES Stephen R. Anderson
LATE ANTIQUITY Gillian Clark
LAW Raymond Wacks
THE LAWS OF THERMODYNAMICS
Peter Atkins
LEADERSHIP Keith Grint
LINCOLN Allen C. Guelzo
LINGUISTICS Peter Matthews
LITERARY THEORY Jonathan Culler
LOCKE John Dunn
LOGIC Graham Priest
LOVE Ronald de Sousa
MACHIAVELLI Quentin Skinner
MADNESS Andrew Scull
MAGIC Owen Davies
MAGNA CARTA Nicholas Vincent
MAGNETISM Stephen Blundell
MALTHUS Donald Winch
MANAGEMENT John Hendry
MAO Delia Davin
MARINE BIOLOGY Philip V. Mladenov
THE MARQUIS DE SADE John Phillips
MARTIN LUTHER Scott H. Hendrix
MARTYRDOM Jolyon Mitchell
MARX Peter Singer
MATERIALS Christopher Hall
MATHEMATICS Timothy Gowers
THE MEANING OF LIFE Terry Eagleton
MEDICAL ETHICS Tony Hope
MEDICAL LAW Charles Foster
MEDIEVAL BRITAIN John Gillingham
and Ralph A. Griffiths
MEMORY Jonathan K. Foster
METAPHYSICS Stephen Mumford
MICHAEL FARADAY
Frank A. J. L. James
MICROBIOLOGY Nicholas P. Money
MICROECONOMICS Avinash Dixit
THE MIDDLE AGES Miri Rubin
MINERALS David Vaughan
MODERN ART David Cottington
MODERN CHINA Rana Mitter
MODERN FRANCE Vanessa R. Schwartz
MODERN IRELAND Senia Pašeta
MODERN JAPAN
Christopher Goto-Jones
MODERN LATIN AMERICAN
LITERATURE
Roberto González Echevarría
MODERN WAR Richard English
MODERNISM Christopher Butler
MOLECULES Philip Ball
THE MONGOLS Morris Rossabi
MORMONISM Richard Lyman Bushman
MUHAMMAD Jonathan A. C. Brown
MULTICULTURALISM Ali Rattansi
MUSIC Nicholas Cook
MYTH Robert A. Segal
THE NAPOLEONIC WARS
Mike Rapport
NATIONALISM Steven Grosby
NELSON MANDELA Elleke Boehmer
NEOLIBERALISM Manfred Steger and
Ravi Roy
NETWORKS Guido Caldarelli and
Michele Catanzaro
THE NEW TESTAMENT
Luke Timothy Johnson
THE NEW TESTAMENT AS
LITERATURE Kyle Keefer
NEWTON Robert Iliffe
NIETZSCHE Michael Tanner

NINETEENTH-CENTURY BRITAIN Christopher Harvie and H. C. G. Matthew
THE NORMAN CONQUEST George Garnett
NORTH AMERICAN INDIANS Theda Perdue and Michael D. Green
NORTHERN IRELAND Marc Mulholland
NOTHING Frank Close
NUCLEAR POWER Maxwell Irvine
NUCLEAR WEAPONS Joseph M. Siracusa
NUMBERS Peter M. Higgins
NUTRITION David A. Bender
OBJECTIVITY Stephen Gaukroger
THE OLD TESTAMENT Michael D. Coogan
THE ORCHESTRA D. Kern Holoman
ORGANIZATIONS Mary Jo Hatch
PAGANISM Owen Davies
THE PALESTINIAN-ISRAELI CONFLICT Martin Bunton
PARTICLE PHYSICS Frank Close
PAUL E. P. Sanders
PEACE Oliver P. Richmond
PENTECOSTALISM William K. Kay
THE PERIODIC TABLE Eric R. Scerri
PHILOSOPHY Edward Craig
PHILOSOPHY OF LAW Raymond Wacks
PHILOSOPHY OF SCIENCE Samir Okasha
PHOTOGRAPHY Steve Edwards
PHYSICAL CHEMISTRY Peter Atkins
PLAGUE Paul Slack
PLANETS David A. Rothery
PLANTS Timothy Walker
PLATO Julia Annas
POLITICAL PHILOSOPHY David Miller
POLITICS Kenneth Minogue
POSTCOLONIALISM Robert Young
POSTMODERNISM Christopher Butler
POSTSTRUCTURALISM Catherine Belsey
PREHISTORY Chris Gosden
PRESOCRATIC PHILOSOPHY Catherine Osborne
PRIVACY Raymond Wacks
PROBABILITY John Haigh
PROGRESSIVISM Walter Nugent
PROTESTANTISM Mark A. Noll
PSYCHIATRY Tom Burns
PSYCHOLOGY Gillian Butler and Freda McManus
PSYCHOTHERAPY Tom Burns and Eva Burns-Lundgren
PURITANISM Francis J. Bremer
THE QUAKERS Pink Dandelion
QUANTUM THEORY John Polkinghorne
RACISM Ali Rattansi
RADIOACTIVITY Claudio Tuniz
RASTAFARI Ennis B. Edmonds
THE REAGAN REVOLUTION Gil Troy
REALITY Jan Westerhoff
THE REFORMATION Peter Marshall
RELATIVITY Russell Stannard
RELIGION IN AMERICA Timothy Beal
THE RENAISSANCE Jerry Brotton
RENAISSANCE ART Geraldine A. Johnson
REVOLUTIONS Jack A. Goldstone
RHETORIC Richard Toye
RISK Baruch Fischhoff and John Kadvany
RITUAL Barry Stephenson
RIVERS Nick Middleton
ROBOTICS Alan Winfield
ROMAN BRITAIN Peter Salway
THE ROMAN EMPIRE Christopher Kelly
THE ROMAN REPUBLIC David M. Gwynn
ROMANTICISM Michael Ferber
ROUSSEAU Robert Wokler
RUSSELL A. C. Grayling
RUSSIAN HISTORY Geoffrey Hosking
RUSSIAN LITERATURE Catriona Kelly
THE RUSSIAN REVOLUTION S. A. Smith
SCHIZOPHRENIA Chris Frith and Eve Johnstone
SCHOPENHAUER Christopher Janaway
SCIENCE AND RELIGION Thomas Dixon
SCIENCE FICTION David Seed
THE SCIENTIFIC REVOLUTION Lawrence M. Principe
SCOTLAND Rab Houston
SEXUALITY Véronique Mottier

SHAKESPEARE Germaine Greer
SIKHISM Eleanor Nesbitt
THE SILK ROAD James A. Millward
SLEEP Steven W. Lockley and Russell G. Foster
SOCIAL AND CULTURAL ANTHROPOLOGY John Monaghan and Peter Just
SOCIALISM Michael Newman
SOCIOLINGUISTICS John Edwards
SOCIOLOGY Steve Bruce
SOCRATES C. C. W. Taylor
THE SOVIET UNION Stephen Lovell
THE SPANISH CIVIL WAR Helen Graham
SPANISH LITERATURE Jo Labanyi
SPINOZA Roger Scruton
SPIRITUALITY Philip Sheldrake
SPORT Mike Cronin
STARS Andrew King
STATISTICS David J. Hand
STEM CELLS Jonathan Slack
STRUCTURAL ENGINEERING David Blockley
STUART BRITAIN John Morrill
SUPERCONDUCTIVITY Stephen Blundell
SYMMETRY Ian Stewart
TEETH Peter S. Ungar
TERRORISM Charles Townshend
THEATRE Marvin Carlson
THEOLOGY David F. Ford
THOMAS AQUINAS Fergus Kerr
THOUGHT Tim Bayne
TIBETAN BUDDHISM Matthew T. Kapstein
TOCQUEVILLE Harvey C. Mansfield
TRAGEDY Adrian Poole
THE TROJAN WAR Eric H. Cline
TRUST Katherine Hawley
THE TUDORS John Guy
TWENTIETH-CENTURY BRITAIN Kenneth O. Morgan
THE UNITED NATIONS Jussi M. Hanhimäki
THE U.S. CONGRESS Donald A. Ritchie
THE U.S. SUPREME COURT Linda Greenhouse
UTOPIANISM Lyman Tower Sargent
THE VIKINGS Julian Richards
VIRUSES Dorothy H. Crawford
WITCHCRAFT Malcolm Gaskill
WITTGENSTEIN A. C. Grayling
WORK Stephen Fineman
WORLD MUSIC Philip Bohlman
THE WORLD TRADE ORGANIZATION Amrita Narlikar
WORLD WAR II Gerhard L. Weinberg
WRITING AND SCRIPT Andrew Robinson

Available soon:

PLATE TECTONICS Peter Molnar
DANTE Peter Hainsworth and David Robey
ANCIENT ASSYRIA Karen Radner
TAXATION Stephen Smith
CORRUPTION Leslie Holmes

Leslie Klenerman

HUMAN ANATOMY

A Very Short Introduction

Great Clarendon Street, Oxford, OX2 6DP,
United Kingdom

Oxford University Press is a department of the University of Oxford. It furthers the University's objective of excellence in research, scholarship, and education by publishing worldwide. Oxford is a registered trade mark of Oxford University Press in the UK and in certain other countries

First edition published in 2015

Impression: 1

Published in the United States of America by Oxford University Press
198 Madison Avenue, New York, NY 10016, United States of America

British Library Cataloguing in Publication Data
Data available

Library of Congress Control Number: 2014949429

ISBN 978-0-19-870737-0

Printed in Great Britain by
Ashford Colour Press Ltd, Gosport, Hampshire

Contents

Preface

This book has been written to provide an introduction to the study of the structure of the human body. This is a complex machine with many different systems which combine to produce smooth function. Our knowledge has been acquired over centuries and still continues to enlarge now, predominantly at the molecular level. It is hoped that it will provide the principles and general structural plan to what at first appears to be a formidably complex subject, with a language of its own.

The text is inevitably not comprehensive and also it does not deal with histology (microscopic anatomy). Nevertheless it is hoped that it will provide a stimulus for further study of the subject and introduction for those about to embark on anatomy at university. Knowledge of the basic structure of the body is helpful for understanding the situation when things go wrong and we all should know about the body we inhabit for the duration of our lives.

The book has been planned to show first how anatomy started. This is followed by chapters on the skeleton and the two major systems the cardiovascular and central nervous system. The rest of the body is covered on a regional basis. The final chapter is on anthropology and deals with the important changes in our bodies which occurred as a result of development of our upright posture.

Acknowledgements

First, I would like to thank Latha Menon for suggesting that I should write this book, I was delighted to have the opportunity, and thanks also for her skilful editorial guidance. Emma Ma gave me much practical advice. Cecilia Brassett, head of University of Cambridge Clinical Anatomy Department, was meticulous in her review of the text. I am also grateful to my fellow senior demonstrators Humphrey Adair and John Fergus for their helpful comments. Emily Evans has made a signal contribution with her excellent drawings.

My wife Naomi and son Paul, Professor of Immunology in the Nuffield Department of Medicine, University of Oxford, were a source of continual help, criticism, and encouragement throughout, while my other son, David, Professor in the Chemistry Department at the University of Cambridge acted as a lay reader, as did Professor Mike Ashby of the Engineering Department at the University of Cambridge.

List of illustrations

All diagrams by Emily Evans

Chapter 1
The development of anatomy

Anatomy is the science of the form and structure of living organisms.

Sir Wilfred Le Gros Clark

The body is one of the most beautiful and amazing machines in the world, filled with intricate devices, cunningly adjusted, wonderfully adapted.

A.V. Hill

According to J.C. Boileau Grant (1866–1973), a famous professor of anatomy at Toronto University, there are few words with a longer history than the word *anatomy*. If we write *anatome* it is then the name that Aristotle (384–322 BC) gave to the science of anatomy 2,300 years ago. He made the first steps towards accurate knowledge of the subject, although it was derived from dissections of small animals. The word means 'cutting up'—the method by which the study of the structure of living things is made possible.

Herophilus of Chalcedon who worked in the city of Alexandria (300 BC) has been called 'The Father of Anatomy'. He was, so Galen assures us, the first to dissect both human and animal bodies in public. He recognized the brain as the central organ of the nervous system and regarded it as the seat of intelligence.

He was the first to grasp the nature of nerves other than those of the special senses, but continued to use the word *neuron* for sinews and ligaments.

Galen (*c.* AD 150–200) was, after Hippocrates (460–370 BC), the most famous Greek physician and often called 'The Father of Medicine', being one of the greatest biologists of all time. His brilliance completely hypnotized the learned men of the Middle Ages by whom he was dubbed the 'Prince of Physicians'. He elaborated a system of pathology which combined the ideas of Hippocrates on the four humours: blood, yellow bile, black bile, and phlegm. Although he was Greek, he worked in Rome during the time of the Emperor Marcus Aurelius. He wrote extensively on the human body. The best presentation of his anatomical knowledge is found in his great works, *On Anatomical Procedure* and *Of the Uses of the Parts of the Body of Man*. He had studied a human skeleton in Alexandria, and he described the varieties of bones and used terms such as *apophysis* and epiphysis, which are still used today to describe an attachment of muscle to a growth site and to the end of a growing bone. Throughout his works the muscles are probably the structures he described most accurately, with frequent references to the form and function of muscles in various animals. Among Galen's most remarkable efforts are the accurate investigations he made of the nervous system. He found that injury to the spinal cord in the neck, between the first and second vertebrae, caused instantaneous death; and that injury to the section between the third and fourth produced cessation of breathing. Below the sixth vertebra, he found that injury gave rise to paralysis of the thoracic muscles, and respiration was carried on only by the diaphragm. If, however, the damage, or lesion, was lower still the paralysis was confined to the lower limbs, bladder, and intestines. After Galen died there was no progress for almost a thousand years, and his writings remained as the standard text throughout the Middle Ages and into the 16th century, until the time of Vesalius. Intellectual leadership passed about the 8th century to people of Arabic speech and remained with them until the 13th century. Thus it came about that the most important documents of Greek medicine were translated into Arabic.

Few disciplines are more surely based on the work of one man than anatomy is on Andreas Vesalius (1514–64) (Figure 1). As Professor of Anatomy in the University of Padua he dissected human cadavers, and in 1543, the same year as the publication of Nicolaus Corpernicus's treatise on the heliocentric theory, Vesalius produced his masterpiece, *On the Fabric of the Human Body*. This monograph is the first great positive achievement of science in modern times. His illustrations of muscles or of bones are presented as part of an image of the whole body of a person, and that person is presented with a background in keeping with scenes from life at that time (Figure 2). He introduced anatomical terms still used today, such as *Atlas* (the first cervical vertebra); *Incus* (ossicle in the ear); and *Mitral Valve* (atrioventricular valve on the left side of heart).

During the Renaissance in the 15th century, artists became interested in the accurate representation of the human body. There is evidence that some such as Albrecht Durer (1471–1528), Michelangelo (1475–1564), and Leonardo da Vinci (1452–1519) used human dissection for this purpose. Leonardo excelled in the study of muscles and produced remarkable diagrams showing their actions. Unfortunately his writings and drawings were lost for nearly 500 years and only became more generally known in the 1900s.

English anatomy was awakened from its dormant state by William Harvey (1578–1657). He was educated at Caius College, Cambridge, and worked in Padua for five years. His great work written in Latin, *An Anatomical Dissertation on the Movement of the Heart and Blood Vessels* was published in 1628. Until then it was thought that blood moved from the right to the left side of the heart through minute perforations in the interventricular septum. After countless experiments on different types of animals, Harvey

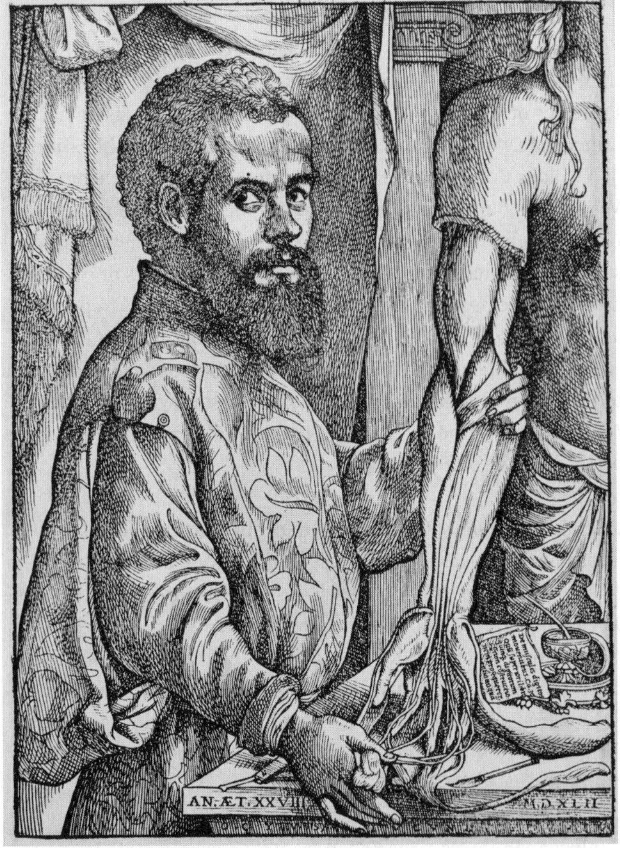

1. Andreas Vesalius (1514–64): a key figure in the development of anatomy.

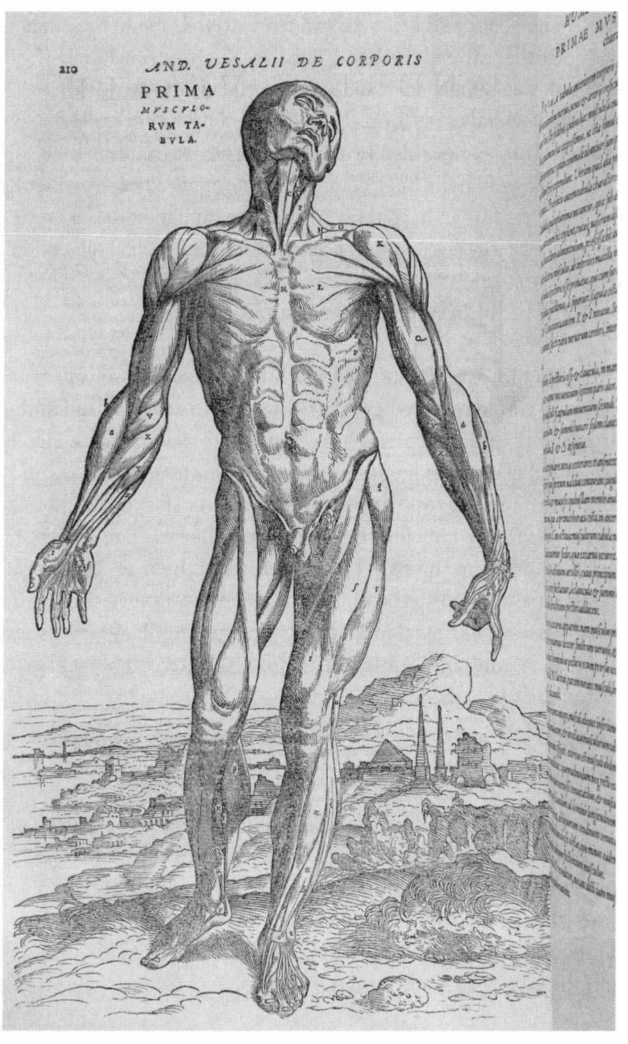

2. Anatomical figure of a human male in Andreas Vesalius' book *On the Fabric of the Human Body*.

finally showed that there were two circulations, one to the lungs and the other around the body. The first English translation of his book was published in 1653. When Harvey's father died in 1623 he decided to perform the autopsy on his father's body himself. He was rewarded by the opportunity to examine his father's exceptionally huge colon, an anatomical curiosity which he would refer to in his public lectures. As human bodies tend to decay after four days without preservation, anatomical dissection required determination and a strong stomach, and winter would generally be the best time for such studies.

Marcello Malpighi (1628–94) was an Italian professor at Bologna who developed great skill in minute observation and supplied the missing element in Harvey's investigations. He described the actual passage of blood from the arteries to the veins through the capillary blood vessels. As Harvey did not use a microscope he knew nothing of the capillaries. The object which yielded up the secret was the lung of the frog. This organ in the frog happens to be almost transparent, is very simple in structure and has particularly conspicuous capillary vessels on its surface. Malpighi could hardly have selected an object better suited for this particular research.

After Vesalius

After Vesalius, anatomy was the queen of the medical sciences for three centuries, giving rise to the disciplines of physiology and pathology. Further progress was made predominantly by practising surgeons. Dissections were carried out as a form of punishment on the bodies of criminals who had been hanged, and these were the only legal dissections.

In 1832 the British parliament passed the Anatomy Act which granted licences to teachers of anatomy and allowed physicians, surgeons, and medical students legal access to corpses that had been unclaimed after death. This put an end to the procurement of parts by body snatching and murders. This Act has recently been updated by the Human Tissue Act of 2004. In the United States and Canada, the Flexner report on medical education in 1910 included dissection of the human body as a requirement of the medical curriculum in universities.

Although anatomy is a basic pillar in the study of medicine, the time devoted to it in the curriculum has gradually diminished because of the competing interests of newer subjects, such as molecular biology, and increased knowledge in areas such as biochemistry, etc. This is a retrograde step and despite the fact that anatomy is best learned by dissection there is a dwindling number of medical schools where this takes place. Models and computerized displays cannot replace the value of time spent on dissection. This book gives an overview of the gross anatomy (that is anatomy visible to the naked eye) of the whole body.

Practical problems

Simple dissection with scalpel and forceps is the time-honoured basic technique for the study of anatomy. It is a physical subject, and by patient dissection and handling of the tissues one learns to study and remember the body in three dimensions. This is supplemented by surface anatomy when one can identify key bony points and relate them to underlying structures, which is essential for clinical examination. Nowadays with the advance of imaging it is easy to view the body in cross-section with x-rays using computerized axial tomography (CAT scanning) and ultrasonography and magnetic resonance imaging (MRI). MRI, in particular, made possible the study of function in the nervous system as well as improving our understanding of neural connections.

The language of anatomy

All terms describing position are given in relation to a standard position (i.e. the *anatomical position*) with the body erect and facing the examiner. The palms face forward and the thumbs are away from the body. In this position the forearm is said to be supinated. When rotated so that the palm faces backward it is pronated.

To indicate cuts or sections made through a cadaver there is a series of planes (Figure 3). The median plane is a vertical plane passing from the front to the back of the body through the midline. Any plane parallel to this is a *sagittal plane.* All vertical planes that pass from side to side and are at right angles to the median plane are *coronal planes.* Planes at right angles to both

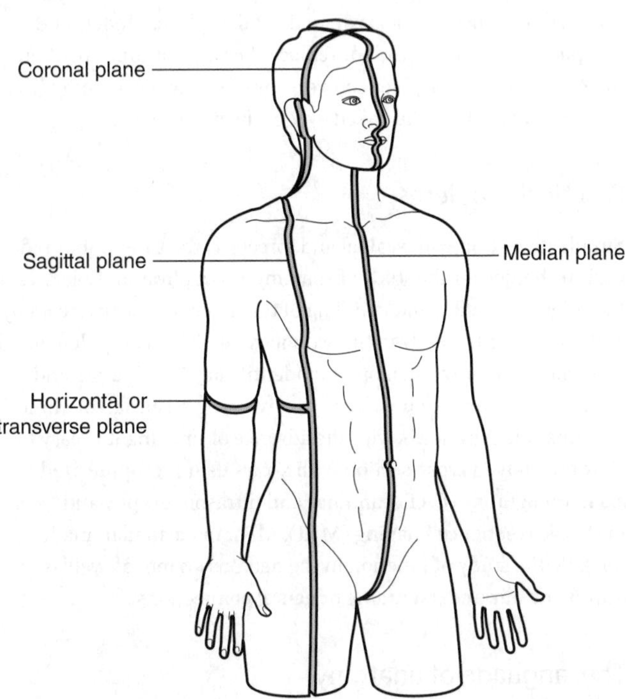

3. The three standard anatomical planes.

coronal and median planes are *horizontal planes*. These planes are also used in descriptions of scans such as the CAT scan (which often provides an image in the horizontal plane—in a cross-section) and the MRI scan (which typically provides a sagittal—or side-on—image).

A further group of terms is used to relate one structure to another using the anatomical position as a reference. A structure may be said to lie above (or *superior*) to another, which lies below (or *inferior*) to the one above it. A structure lying in front of another is *anterior* to it and a structure lying behind another is *posterior* to it. The terms *ventral* and *dorsal* have almost the same meaning as anterior and posterior but are more often applied to surfaces. Two structures can be related to the midline of the body: the one nearest to the midline is described as *medial*; and the one farther away is *lateral*. The terms *proximal* and *distal* are useful when describing parts of structures nearer to or farther away from their origin or from the trunk.

With regard to movements at joints: to *flex* is to bend or make an angle; to *extend* is to stretch or to straighten; to *abduct* is to draw away laterally from the median plane of the body; to *adduct* is the opposite movement in the same plane. Movements of flexion, extension, abduction, and adduction take place, for example, at the wrist joint. To *circumduct* is to perform the movements of flexion, abduction, extension, and adduction in sequence. To *rotate* is to turn or revolve on a long axis medially or laterally. In the foot, *inversion* is elevation of the medial border of the foot; and *eversion* is elevation of the lateral border of the foot.

The names of structures often originate from the language of ancient anatomists around the globe. We still use names and terms with origins in Greek, Latin, Arabic, French, and Anglo-Saxon.

Anatomical variations

Textbooks of anatomy usually describe the most common form in which structures are found in the body, but after a short time in the dissection hall it soon becomes clear that variations from the textbook description are frequently encountered. The majority of these variations are perfectly compatible with normal life and go unrecognized until discovered in the cadaver. For example the division of the sciatic nerve into tibial and fibular components may occur high in the buttock or low in the midthigh. Similarly the division of the brachial artery into radial and ulnar arteries may occur in the arm or at the elbow. Other common examples are the palmaris longus in the forearm and plantaris in the leg. One is absent in about 10 per cent of limbs and the other in 6 per cent. A rare (one in 7,000) but interesting variation is *situs inversus*, a complete reversal of asymmetry in all organs usually with normal physiology. *Dextrocardia*, in which the heart lies on the right side of the thorax instead of the left, forms part of the picture. Human variations are a source of interest because they provide insight into developmental anatomy.

Chapter 2
The skeleton and its attachments

It is notorious that man is constructed on the same general type or model with other mammals. All the bones in his skeleton can be compared with corresponding bones in a monkey, bat or seal. So it is with his muscles, nerves, blood vessels and internal viscera.

Charles Darwin, *The Descent of Man*

The skeleton gives the vertebrate body shape, supports its weight, and protects soft parts such as nerves and blood vessels. It also offers a system of levers that, together with muscles, produces movement. Since mineralized parts of the skeleton often survive fossilization better than do soft tissues, our most direct contact with long-extinct animals is through their skeletons. The story of vertebrate evolution is written in the architecture of the skeleton. This chapter describes the skeleton and the structures attached to it—starting from the deepest layer (bone) to the most superficial (skin).

The skeleton must be robust yet light and elastic, thus allowing mobility while maintaining strength. Bones are classified according to their shape and appearance into, short, long, flat, and irregular bones. Some are preformed in the embryo as thickened layers of so-called 'connective' tissue, which in most cases develop into a soft cartilage before full bone formation ('ossification'). In some cases, when there is no preformed cartilage replica, bones such as the collar bone (clavicle) may develop in membrane.

Cartilage is less hard than bone because it lacks calcium crystals and it is flexible. It is familiar to fish eaters as the 'bones' of the fish skeleton. Most of our skeleton starts as cartilaginous bones, in the limbs, trunk, and base of skull. The site at which bone first appears is known as the primary centre of ossification. Before this is complete, one or more secondary centres of ossification typically appear at the upper and lower ends of the bone after birth. These regions are termed the *epiphyses* and are separated from the main shaft of the bone (*diaphysis*) by a thin plate of cartilage (epiphyseal plate). That part of the bone adjacent to this plate is a site of intense growth (metaphysis) and is well served with blood vessels. As a result it is also a site of weakness where fractures occur until the plate has stopped growing and becomes fixed or 'fused'. The growing end of a bone refers to the end that contributes most to bone growth. The growing ends of limb bones become of particular importance in the treatment of unequal growth in children. Equalization of limb length can be achieved by surgically limiting the expansion of the more rapidly growing cartilage with staples placed across it (Figure 4).

The skeleton also serves as a mineral reservoir and plays an important role in calcium and phosphate balance. Bone is constantly renewed by two opposite activities, the breakdown of old bone and the formation of new bone. The bulk of the organic material in bone (about 35 per cent) is made up of a single protein, known as *collagen*. This major structural protein together with calcium hydroxyapatite gives bone its strength. Hydroxyapatite is composed of calcium phosphate in repeating units with the chemical formula $Ca_{10}(PO4)_6(OH)_2$. A bone typically has a dense outer (cortical) layer of bone and an inner region of spongy bone. The latter contains a network of struts and braces for transmission and distribution of stresses. It also contains within a cavity the bone marrow—a soft material

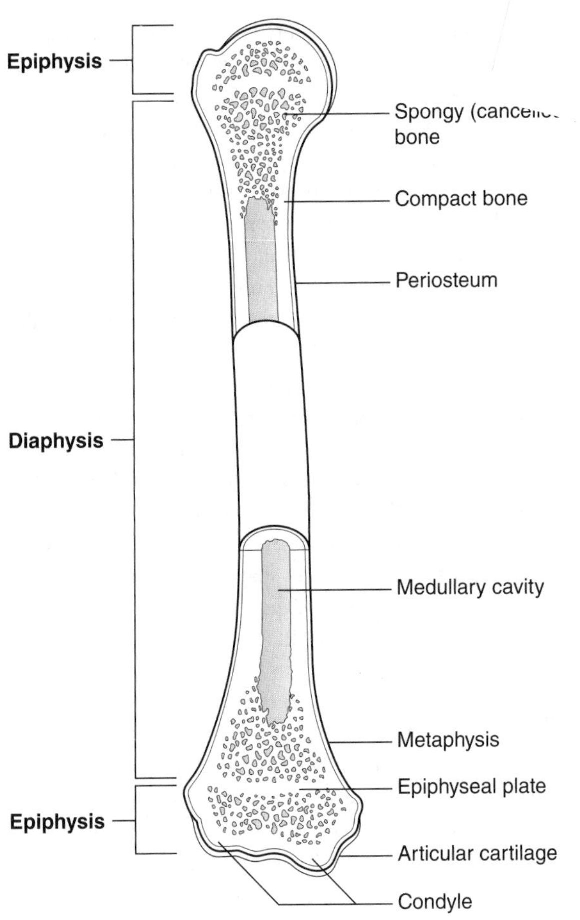

4. A long bone.

consisting of fat and vascular tissue, where the components of blood develop. Research has shown that the long hollow bones of mammals are optimized for maximum strength-to-weight ratio. The optimum cross-section for a long bone is a circle with an internal radius that is about 60 per cent of the external radius.

The vascular supply of bones is obtained from neighbouring arteries; and in long bones many small vessels enter near the ends. Their tracks can be seen as small holes in dry bones. Larger vessels known as 'nutrient arteries' enter the shaft. Other smaller blood vessels enter the bone through the outer covering layer (periosteum). Autonomic nerves (see Chapter 4) pass into the bone with the blood vessels. This means that removing the periosteum of a long bone may result in bone death. However, this does not apply to the bones of the vault of the skull as their blood supply is almost completely from the inner surface of the skull.

Bone responds to mechanical forces and realigns according to stress and strain. This adaptability is responsible for the hypertrophied (thickened) bones of the playing arm of a tennis player and the spontaneous realignment with growth of fractures in children. Similarly there is atrophy (decrease in mass) of bone and loss of bone strength following prolonged bed rest or exposure to microgravity as in astronauts in space. This principle of adaptability is known as *Wolff's Law*, and bears the name of one of its early proponents Julius Wolff (1836–1902).

Cartilage

Cartilage or gristle is a versatile type of connective tissue, which in addition to collagen contains a firm gel-like substance (derived from complex sugars). As a result it can bear weight without bending but retains flexibility. There are three types of cartilage: hyaline, fibrocartilage, and elastic. Hyaline cartilage (from the Greek *hualos*, a transparent stone) is white, resilient, and has the potential to form bone. In fact all the bones, except certain skull bones and the clavicle, were preformed in hyaline cartilage in early life. It persists in the adult as the cartilage in most joints; as costal cartilage at the ends of the ribs joining the breastbone (sternum); and as supportive cartilages of the nose and respiratory tract. Fibrocartilage develops wherever fibrous tissue is subjected to great pressure and is thus particularly strong. It occurs in intervertebral discs, and the semilunar cartilages of the knee (menisci). Elastic cartilage, as the name suggests, is pervaded by elastic fibres and has a yellow appearance. It is found in the external ear and mobile cartilages such as the epiglottis, which guards the entrance to the larynx.

Together, these two tissues—bone and cartilage—have evolved over time to create the skeleton. As ancient ancestral vertebrates evolved from passive filter feeders to active predators, they also became larger with increased mobility. The development of the internal skeleton supported larger muscles and increased movement while protecting internal organs. This is reflected in the human skeleton, which can be divided into two major parts: the axial (or midline) skeleton, consisting of the skull, spine, and bones of the chest (which have a largely protective role); and the appendicular skeleton, which consists of the bones of the upper and lower limbs driving locomotion (Figure 5).

The axial skeleton—spine and ribs

The human vertebral column shows special adaptations associated with a habitual erect posture and a unique bipedal gait. Instead of acting as a suspension bridge between forelimbs and hind limbs as in a quadruped, it is required to support the weight above the pelvic girdle. The progressively greater load born by individual bones is reflected by their increasing size from the skull to the sacrum, similar to the mast of a sailing vessel with its base in the hull. It is composed of thirty-three vertebrae joined by intervertebral discs made of cartilage: the upper seven are named cervical (neck) vertebrae, the next twelve are thoracic (chest) vertebrae; below these are five large lumbar (lower back) vertebrae, which are followed by five vertebrae that form the sacrum fused together within the pelvis. Galen regarded the sacrum as the most important bone of the spine, hence its name. A variable number (three to five) of fused coccygeal vertebrae—remnants of our long lost tail—completes

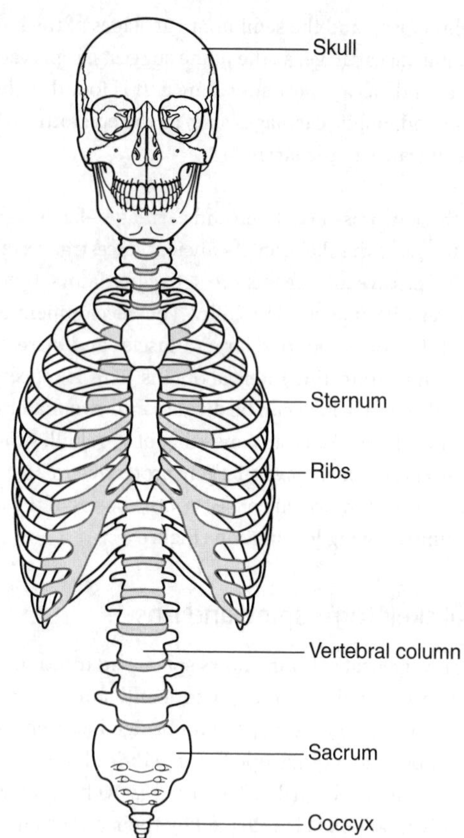

5. The axial skeleton.

the vertebral column. The sacrum is attached by strong interosseous ligaments to the pelvis at the sacroiliac joints and thus body weight is transmitted from the head to the lower limbs via the hip joints. The centre of gravity of the body has been calculated to be located around the second sacral piece. Through verterbrate evolution, typically a region elongates or shortens simply by changing the lengths of each individual vertebra rather than adding more bones—thus the neck of a giraffe or an owl each will still have seven cervical vertebrae.

The vertebrae all have the same basic structure. There is a solid cylindrical body providing strength and, protecting the spinal cord, a dorsal neural arch which is formed by two flat sheets (laminae) joined at the rear. The dorsal arch is joined to the main vertebral body by two short pieces of bone (pedicles) (Figure 6a). To allow muscle attachments and thus movement of the column there are small spikes of bone (transverse processes) on each side of the vertebral body and a larger spinous process at the rear. This arrangement thus combines protection, flexibility, and strength.

Before birth the vertebral column has a single primary curvature, concave anteriorly. A cervical curvature (convex anteriorly in the neck) begins to appear when the child starts to sit and a similar curvature appears in the lumbar region when he or she stands. Changes occur in the spine with age. An increase of the curvature

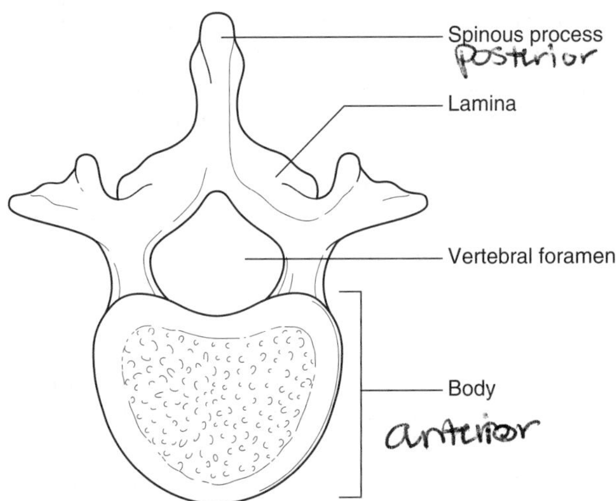

6a. The parts of a typical vertebra.

of the thoracic region, often seen in elderly women (the so-called dowager's hump), occurs due to osteoporosis (bone thinning) with bulging of the discs into the soft vertebrae. Likewise, exaggeration of the anterior curve of the lumbar spine occurs in pregnancy due to the bulk of the developing child in the abdominal cavity. An abnormal lateral curvature, which is typically accompanied by some rotation, is called *scoliosis*. This occurs as a progressive condition in some girls around puberty in the midthoracic region, and is called *idiopathic* scoliosis because no known cause has yet been discovered. However, during normal development, our vertebral column is so well-balanced that, when standing, most of the muscles of the human body are almost completely inactive. This is one of the most striking ways in which humans are more efficient machines than the vast majority of mammals.

The critical structures joining the bodies of the vertebrae are the intervertebral discs. Between the disc and the vertebral body lies a thin layer of hyaline cartilage through which nutrients pass. The disc consists of an outer ring of layers of fibrous tissue and fibrocartilage called the *annulus fibrosis* and a soft core, the *nucleus pulposus*. The design of the discs provides a strong union and shock absorption. More diffusion of water and nutrients occurs at night when sleeping and not bearing weight than during the day. By very accurate measurements it can be shown that one is tallest early in the morning and shortest at night after a day of weight-bearing activity. Astronauts in space, experiencing microgravity, become longer due to the expansion of discs, and they require special, reinforced suits to compress the vertebral column. Discs are relatively avascular and the lumbar discs, the last two in particular, which bear the highest loads are prone to degeneration from the mid-20s onwards (Figure 6b). Sometimes part of the nucleus of the disc may protrude from beneath the annulus posteriorly. This damage causes an inflammatory reaction and compression of the adjacent nerve root which results in pain in the thigh and leg known as sciatica, or 'a slipped disc'.

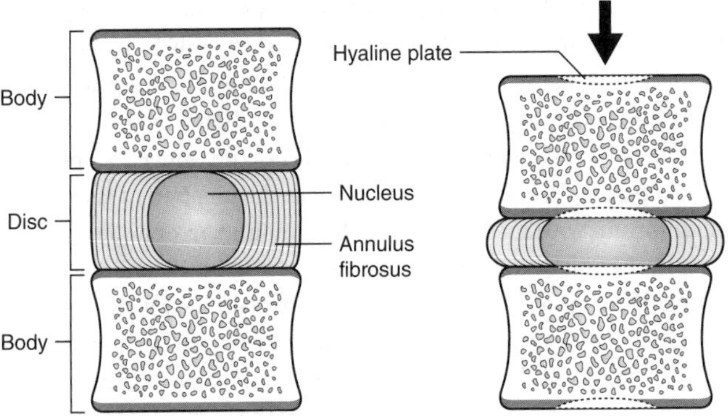

6b. An intervertebral disc when loaded.

The verterbral column is also connected by a set of ligaments, essential for stability and movement. One in particular, the nuchal ligament, may have evolved independently in man (see Chapter 8) and other animals that are well adapted for running such as dogs and horses. It bridges the gap between the base of the skull and the spinous process of the seventh cervical vertebra in the neck. It is also found in animals with massive heads such as giraffes and elephants. Small movements occurring between a number of vertebrae can together produce an appreciable total range. The range of movement in each section of the spine is determined by the alignment of the articular processes. Forward and sideways movements occur in the cervical and lumbar spine, but rotation is confined to the atlanto-axial joint between the atlas and the axis (first and second cervical vertebrae) and the thoracic spine. Overall the neck is the most mobile and most vulnerable region of the spine.

The thoracic cage is formed by the vertebral column behind, the ribs on either side, and the sternum in front. At the front of the chest, the first seven ribs are connected by costal cartilages to the sternum (true ribs). The cartilages of the eighth, ninth, and tenth join with the cartilage of the rib above and are called 'false ribs', while the last two ribs, which are free to move anteriorly, are known as 'floating ribs'. The sternum consists of three parts: the manubrium (named after a sword handle, with the sternum as the blade) is the uppermost part and forms joints on both sides with the clavicles, the first costal cartilages and upper part of the second costal cartilages. The joint at the junction of the manubrium with the body of the sternum is at an oblique angle (angle of Louis) which is a useful clinical landmark as it is easily palpable and is at the level of the disc between the fourth and fifth thoracic vertebrae. It is also helpful as a landmark for localization of specific ribs and thus the dimensions of the heart during a clinical examination. At the lower end of the sternum is the xiphoid process, which usually remains cartilaginous well into adult life. The sternum offers a site of origin for chest muscles. In birds a keel or crest projects from the sternum in order to afford the powerful wing muscles an increased surface of origin. The skull is dealt with in Chapter 4.

The appendicular skeleton

The upper and lower limbs possess many similarities. Each is supported on a girdle linked to the axial skeleton. Each has a single proximal long bone and a pair of distal long bones, which articulate with a five-digit hand or foot. With the development of bipedal gait the upper limb has been freed from the task of weight bearing and has become adapted for investigatory and manipulative skills. In contrast to the foot, there have not been changes to the bony structure of the hand, which remains the basic pentadactyl structure found in all mammals Figure 7a.

The skill and control associated with the hand are primarily the result of developments that took place in the brain with the adoption of bipedal gait. A balanced posture allowed the development of tool-using and -making, which in turn led to development of the brain through a positive feedback mechanism.

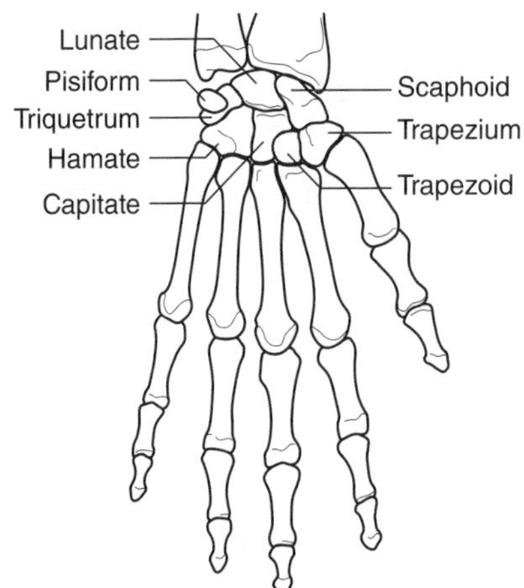

7a. The bones of the hand.

The upper limb is suspended from the bony pectoral girdle, which links it to the trunk. The main features of the skeleton of the upper limb are illustrated in Figure 7b. The upper limb is moored to the head, neck, and trunk by muscles which may be likened to guy ropes. The clavicle (collar bone) is the only skeletal connection between the scapula (shoulder blade) and the rib cage. The clavicle articulates near the midline with the manubrium of the sternum and laterally with the acromion of the scapula to form the shallow sternoclavicular and acromioclavicular joints, respectively. The name *clavicle* is derived from the Latin word for a little key, as a Roman key was S-shaped. Animals such as the dog, ox, or horse that use their forelimbs merely for locomotion (i.e. forward and backward motions) have either no clavicles or only very rudimentary ones, while primates, rodents, guinea pigs, and bats, which employ their forelimbs for climbing, burrowing, or flying, have well-developed clavicles, which allow for the required side-to-side movements.

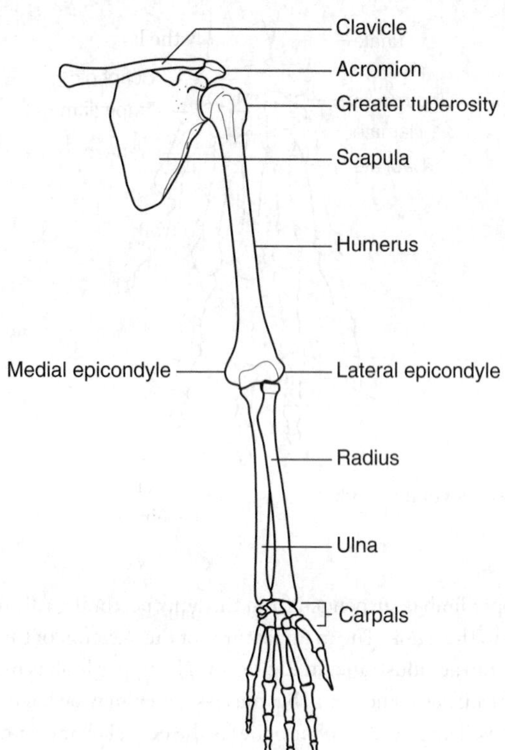

7b. The bones of the upper limb.

The scapula or shoulder blade is a flat, triangular bone, which is attached to the thorax and vertebral column by a number of stabilizing muscles; it provides a platform from which the arm can hang from the shoulder. In the resting position, the medial border lies parallel to the vertebral column. At the back a horizontal ridge of bone known as the *spine* divides the scapula into two large depressions, the supraspinous and infraspinous fossae, where muscles attach. The acromion is the prolongation of the spine and meets with the lateral end of the clavicle to form the small but crucial acromioclavicular joint. Laterally the large shallow socket (glenoid fossa) allows movement with the head of the humerus to form the mobile shoulder joint.

The humerus is the long bone of the arm that articulates with the scapula above and the bones of the forearm below at the elbow. The 'anatomical' neck separates the head from the greater and lesser tuberosities (or tubercles)—protrusions to which important muscles can attach for arm movement. The 'surgical' neck, which separates the proximal end from the humeral shaft, is so-called as it is a common site of fractures in the elderly. At the elbow joint there is a notch-like pulley for the ulna (trochlear) and a round prominence (capitulum) for the head of the radius which is very suitable for rotation. The radius and ulna are bound to each other by ligaments and act as a team. An interosseous membrane connects them throughout the length of their shafts. They articulate with each other at the superior and inferior radio-ulnar joints which are responsible for rotation of the forearm—supination meaning palm upward and pronation, palm down.

The radius articulates with the wrist (carpus), which consists of eight small bones arranged in two rows. Minimal movement normally occurs between these bones, which are tightly bound to each other by tough ligaments. Injury to these ligaments may result in instability of the wrist. The carpal bones are arranged in two rows, each of four bones. In the proximal row (adjacent to the radius) are the scaphoid, lunate, triquetrum, and pisiform, while the second row consists of the trapezium, trapezoid, capitate, and hamate. Two of these bones are commonly injured during a fall on an outstretched hand. The most common of these is the scaphoid, which has the added complication of an unusual blood supply, which enters from its distal end, and is thus likely to be damaged, with impaired healing. It can be felt in the so-called 'anatomical snuff box', which lies at the base of the thumb; tenderness at this site after a fall may point a clinician towards a scaphoid fracture.

The five metacarpals are long bones that form the main skeleton of the hand. Little movement occurs between the bones of the wrist and the metacarpals except at the saddle-shaped first carpometacarpal joint, which allows free movements of the thumb. The Primate family all show some degree of functional independence of the thumb, however because of its shortness they are unable to perform the movement of *opposition* by which we are able to place the pulpy surface of the thumb squarely in contact with the terminal pads of the fingers. This movement is by far the most effective for the handling of small and delicate objects. Isaac Newton is said to have remarked that in the absence of any other proof, the thumb alone would convince him of God's existence. There are three phalanges (from the Greek for military formation with soldiers in parallel lines) in each finger but only two in the thumb (and great toe).

The skeleton of the pelvic girdle and lower limbs is shown in Figures 8a and 8b. The weight of the trunk can be seen to be transmitted from the sacrum, through the almost immobile sacroiliac joints to the pelvic bones, and from these through the stable but mobile hip joints to the thigh bone (femur) on each side. Each femur (the longest bone in the body) is directed downwards and inwards towards the knee joint, where it articulates with the tibia (but not the fibula) and the patella (kneecap). The slender fibula only provides muscle attachment in its upper part, while its lower end is firmly bound to the tibia to form part of the ankle joint. It does not bear weight, which is the role of the tibia. The line of gravity from the head passes through the axis bone (the second cervical vertebra), just in front of the sacrum, behind the centres of the hip joints and in front of the knee and ankle joints. In order for the body to remain balanced in the upright position, the centre of gravity must lie within an area delimited by the heels and balls of the feet, in other words the points at which the front of the foot touches the ground—a point worth remembering when skiing.

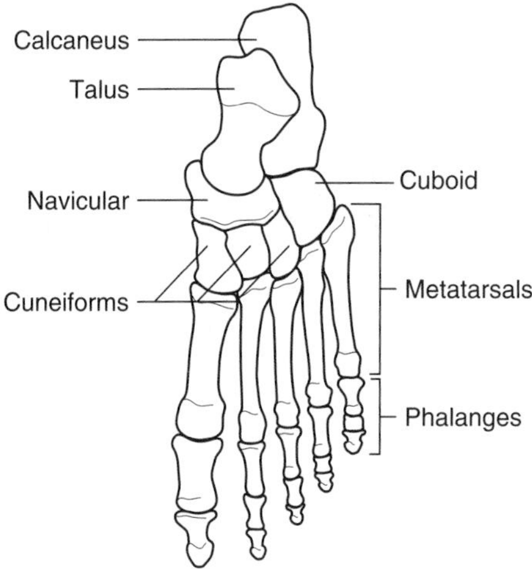

8a. The bones of the foot.

The hip bone (innominate bone) is actually formed by the fusion of three bones that meet at the acetabulum (the socket for the hip; the name derived from the Latin for 'vinegar cup'). The *ilium* forms the large wing of bone that supports the inferior part of the abdominal cavity. The two pubic bones are located in the front and meet in the midline at the pubic symphysis. The ischium is the large bone at the back of the pelvis. The thickened lower portion of the ischium, the *ischial* tuberosity, supports the body weight when sitting and is protected by a bursa (membranous sac containing synovial fluid to reduce friction, which acts like a water cushion).

The femur is the longest and strongest bone in the body and is about one-quarter of the body height. In the kangaroo, a hopping animal, the femur is shorter than the tibia. The upper end has a head, neck, and greater and lesser trochanters, where muscles and ligaments can form strong attachments. The head forms

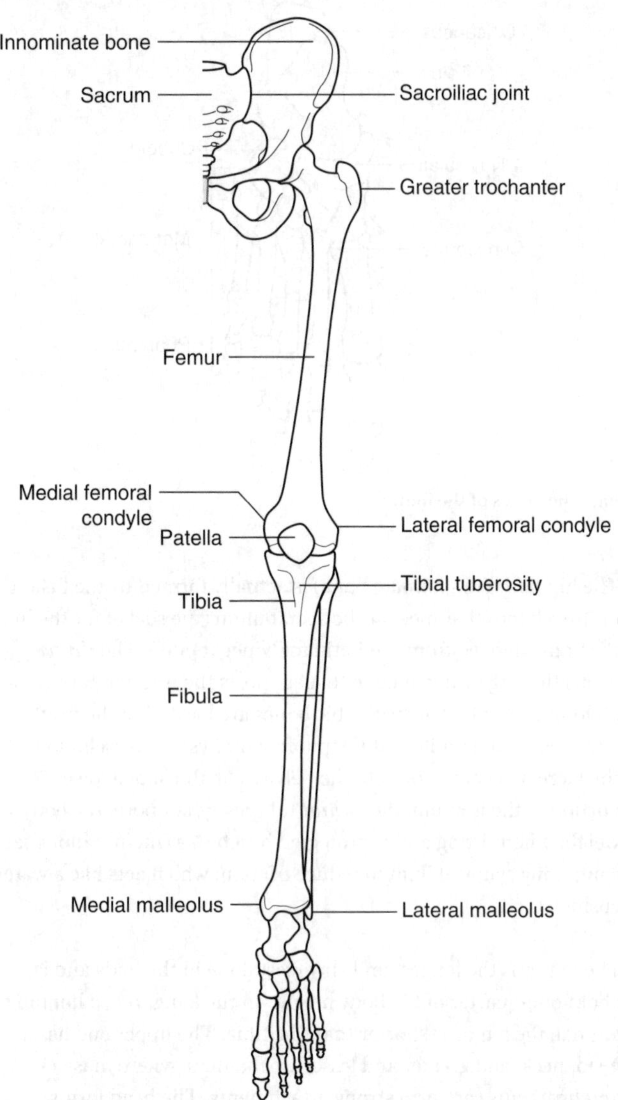

8b. The bones of the lower limb including half of the pelvis.

two-thirds of a sphere, in contrast to the humeral head, which is only one-third, and thus sits much more securely in its socket than the head of the humerus. The greater trochanter at the base of the neck was named by Galen. It is the site for muscle attachments that rotate the hip joint. Similarly the lesser trochanter, a smaller prominence at the upper end of the shaft, is the site of insertion of the iliopsoas muscle, a powerful flexor of the hip (the muscle in cattle that provides fillet steak). The iliopsoas muscle is attached at its upper end to the spine. In regions where tuberculosis is still a common cause of spinal disease (as it was in Western countries in the past), infection in the back, which may be clinically silent, can track down the iliopsoas muscle and may present first as an abscess in the front of the thigh.

The lower end of the femur is divided into two large knuckles, the so-called medial and lateral condyles, which are readily felt. These are covered with hyaline cartilage below and behind for articulation with the tibia. The cartilage from both sides unite in front to form a V-shaped pulley, the patellar surface or trochlea in which the patella slides. The patella (kneecap) is the largest sesamoid bone in the body. Sesamoid bones develop in certain tendons where they rub on convex bony surfaces ('sesamoid' comes from the Arabic for 'like a seed'). The patella develops in the tendon of the major muscle group of the thigh the Quadriceps Femoris. Its free surface is covered with hyaline cartilage, with the remainder of the bone buried in the tendon. Nevertheless, despite all these attachments, the patella can be dislocated in a lateral direction due to injury. Its function is to increase the lever arm of the quadriceps.

The tibia (shin bone; the name is derived from the flute or pipe with the upper end the trumpet shaped mouth and the lower end, the mouth piece) is the large weight-bearing bone of the leg, which connects with the femur above it. The flattened proximal end, the *tibial plateau*, comprises the medial and lateral condyles which articulate with the corresponding femoral condyles. At the front is the prominent tibial tuberosity into which the quadriceps tendon inserts via the patellar ligament, allowing straightening of the leg. Below the tibial plateau the bone narrows to form a shaft with a sharp anterior border, which is easily felt under the skin and is a common site for open fractures, while above the ankle, the shaft splays out to form a prominent protrusion on the inside of the ankle (the medial malleolus). The fibula is slim in comparison to the tibia. It gives origin to muscles and forms a crucial part of the ankle joint. It is moored to the tibia at its upper and lower ends at two tibiofibular joints and linked to the tibia by an interosseous membrane—an arrangement very similar to that of the radius and ulna in the arm. The shaft ends with a splaying of the bone at the distal end to form the lateral malleolus, which forms the lateral buttress of the ankle joint.

The tarsus (instep) is complex and contains seven bones: the calcaneus (heel bone), talus, cuboid, navicular, and three cuneiforms (medial, intermediate, and lateral). The calcaneus is the largest tarsal bone—the anterior two-thirds support the talus while the posterior one-third forms the prominence of the heel and rests on the ground. The joints within the ankle provide both stability and movement. The interactions between three of the critical bones—the talus, calcaneus, and navicular—allow rotation of the forefoot. These movements make it possible to adapt the sole of the foot to the surface on which it is planted. Because of its importance in weight bearing the talus has become the densest bone in the body with very tightly supporting bony strands (trabeculae) forming a strong framework. The midfoot is formed by the articulations of the navicular, cuboid, and three cuneiform bones, while the metatarsals and phalanges are similar in arrangement to those of the hand.

When things go wrong

A number of problems may arise in the development of the skeleton that may require specialist treatment. Variations of spina bifida are the most common of congenital malformations in which the two-halves of the posterior vertebral arch (or several arches) have failed to fuse. Isolated laminar defects can be ignored but more severe defects can be associated with maldevelopment of the spinal cord. About one child in every 250 births has one of these deformities, and many will be stillborn.

Craniocleido dystosis is a disorder in which the most obvious features are poorly formed or absent clavicles together with the absence of muscles attached to them. The shoulders can touch in front of the body. This is the only condition in which growth of the membrane bones alone is affected. It is rare, and occurs once per 200,000 live births. In contrast, achondroplasia, is less rare, with an incidence of from 1:15,000 to 1:40,000 live births. It is a hereditary, congenital familial disturbance in bones developed from cartilage, which results in a peculiar type of dwarfism. This condition is often found in circus dwarves. There is evidence of the existence of achondroplastic dwarves as far back as antiquity. The head is apt to be enlarged because there is failure of growth of the bones of the base of the skull that appear in cartilage and relative overgrowth of the vault of the skull that develops in membrane. There is less inhibition of trunk growth than that of the extremities.

Joints

The site at which two bones are linked together is called a joint or articulation. They can be classified according to their structure and by the type of movement they allow.

Not all joints have a wide range of movement. In the fixed joints of the skull, once growth is complete the separate bone plates are securely connected by interlocking fibrous tissue at sutures, which may ultimately become bone. Semi-moveable joints are linked by cartilage, as for example between vertebrae and the junction of the two hip bones at the pubic symphysis.

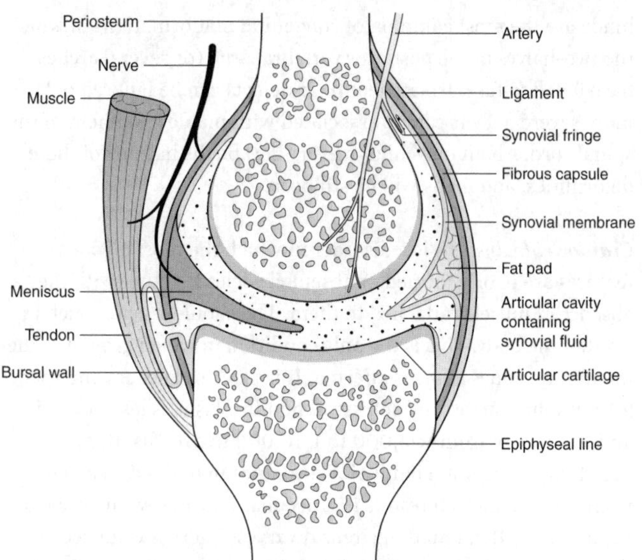

9. The typical structure of a synovial joint.

The most numerous, versatile, and freely moving joints are synovial joints, which exist even between the minute ossicles of the ear. Bones at a synovial joint are separated by a cavity that is enclosed by synovial membrane (synovium) containing a film of lubricating synovial fluid, a filtrate from blood (Figure 9). The bone ends are covered by hyaline cartilage, which is nourished by the synovial fluid. Surrounding the joint is a fibrous capsule. Fibrous thickenings of the capsule called ligaments are anchored to the bones at each end to prevent excessive movement. Muscles around the joint provide stability and produce movement. The amazing feature of synovial joints is the low friction between the hyaline cartilage surfaces, which is three times better than ice sliding on ice. This is based on studies by Sir John Charnley FRS, a pioneer in the development of artificial joints or arthroplasties. Low friction has been the most difficult feature to imitate. At present the best combination is stainless steel on ultra-high molecular weight polyethylene which forms the basis of most joint replacements.

Persistent generalized joint hypermobility (so-called double joints) occurs in about 5 per cent of people. The knees and elbows can be extended beyond 180 degrees (hyperextended) and the hands and feet can attain unusual positions. Hypermobile joints are not necessarily unstable as seen in the performances of acrobats, but they are associated with a tendency to recurrent dislocations of the patella or shoulder.

Muscle

There are three types of muscle. Skeletal or voluntary muscles such as those of the limbs, body wall, and face make up 40 per cent of the weight of a human body in reasonable physical condition. They move human, ant, and elephant alike, so that only a trained eye can see the subtle differences when viewed under a microscope. Cardiac muscle is confined to the heart and smooth, visceral, or involuntary muscle is found in the stomach, intestines, bladder, and blood vessels. Skeletal muscles are under conscious control, in contrast to heart muscle and smooth muscle. The function of the three types of muscle is to contract (and then relax). Skeletal muscle is capable of rapid contraction and smooth muscle of slow sustained contraction without fatigue. Several skeletal muscles may share a common function, or different parts of one muscle may have different functions.

Muscles attach at their ends to the bones. They move through so-called origins and insertions. The fleshy part of a muscle in between is referred to as the *belly*. Most muscles are fibrous at one end, which if rounded is called a tendon. Tendons have the same structure as ligaments but anchor skeletal muscle to bone. When the insertion is flattened and membranous it is called an aponeurosis, as in the flat muscles of the anterior abdominal wall. The name suggests a nervous component, because the ancients did not distinguish between nerves, ligaments, and tendinous structures.

The extent to which a muscle can shorten is about 50 per cent of resting length. To achieve this, muscles display different arrangements of fibres depending on their functions. Muscles that must contract over a great distance need long parallel fibres. If muscle fibres insert more obliquely into the tendon it is possible to include more fibres but shorter ones. Such pennate (feather-like) muscles have greater potential power but contract over shorter distances. Energy may also be stored by tendons. The elasticity of the tendons of weight-supporting limbs means that the strain energy which is produced by muscular contraction and stretching of tendons may be temporarily stored and then used for propulsion as in a spring. The ostrich, the fastest bird on land, makes substantial savings of energy in stretched tendons. They can store twice as much energy per step as humans. This is related to the presence of a number of muscles in their legs with short muscle fibres and long tendons. To the extent that tendons do stretch, they make a splendid elastic material, giving back about 93 per cent of the work put in—about the same as rubber.

An important relationship between joints, muscle, and the skin overlying joints was observed by John Hilton (1805–78), a surgeon anatomist at Guy's Hospital: 'The same trunks of nerves, whose branches supply the groups of muscle moving a joint, furnish also a distribution to the skin over the insertions of the same muscles; and the interior of the join receives its nerves from the same source'. Hilton realized that there must be some reason behind the fact that all these structures were supplied by branches of the same nerve. He came to appreciate that where the joint was inflamed the nerves were stimulated and not only gave referred pain to the skin over the joint, but also caused the muscles controlling the limb to go into spasm, thereby, through immobilization, protecting the joint from further injury. It also explains why disease in the hip may first be felt in the knee, as both joints are supplied by the same nerves (femoral, sciatic, and obturator).

Beneath the skin the various structures that make up the body are held together by 'connective' tissue. Because this tissue is found in some regions in the form of bands it is known as *fascia* from the Latin for 'band'. *Deep fascia* (the third layer of the body wall) is most obvious in the limbs and neck where it is wrapped around the muscles, vessels, and nerves like a bandage (Figure 9). The deep fascia sends intermuscular septa (partitions) between various muscles and groups of muscles and usually blends with or attaches to periosteum (the outer lining of bone). *The superficial fascia* (second layer) is the fat-containing subcutaneous tissue that underlies the greater part of the skin of the body—functioning as an insulating layer. In the face and neck it contains striated muscles that are attached to the skin and control facial expression.

The skin

The skin is the largest organ of the body. It is not merely an envelope, wrapped around our bodies like paper around a parcel, but it is in fact one of our most versatile organs. As it is waterproof it prevents the evaporation and escape of tissue fluids. It becomes thick where it is subject to rough treatment. It is tightly fastened down where it is most likely to be torn off as in the palm and on the sole. It has friction ridges on the finger tips where it is most likely to slip. As an organ it is the regulator of body temperature; it is an excretory organ capable of relieving the kidneys in time of need. It is the factory for the manufacture of vitamin D formed by the action of the ultraviolet rays of the sun on the sterols of the skin, which is necessary for the mineralization of bones and teeth. It has an immense array of sensory receptors to monitor our surroundings and produce appropriate responses. The anatomist Frederick Wood Jones suggested that we should regard the skin as the greatest and most ancient sense organ of the body.

The skin has two parts—the outer layer is the epidermis and beneath it lies the dermis. The cells of the epidermis are flat ('squamous'). The deepest layer of epidermal cells is actively dividing, whereas the outermost layers are dead or dying. The whole superficial surface of the skin is rapidly replaced approximately every 46–8 days. In the deepest layer of the epidermis is a set of specialized cells called melanocytes, which produce a black pigment, melanin, that contributes to the colour of the skin and hair. The dermis, which lies beneath the epidermis, is composed of a dense feltwork of collagen fibres, which in cowhide form the basis of leather. Invaginations of the epidermis are modified to form sweat glands lying in the dermis. Sebaceous glands are present throughout the skin, except for the palms and soles. They produce sebum which lubricates the skin and protects hair. Collagen in the skin creates tension lines (or Langer's lines) first described by Karl Langer (1819–87) an Austrian anatomist. When possible surgeons use skin incisions which lie parallel to Langer's lines as they gape less, providing a better cosmetic effect after the wound has healed.

Chapter 3
The vital systems

The most important feature of the circulation that must always be kept in mind is that is a continuous circuit. That is if a given amount of blood is pumped by the heart, this same amount must also flow through each subdivision of the circulation.

Arthur C. Guyton, Physiologist

The heart

The heart is the most important muscle in the body. Its function is essential for life. It works as a pulsatile, four-chamber pump composed of two atria and two ventricles. The atria act mainly as entrances to the ventricles, but also pump weakly to help move the blood into the ventricles. The ventricles supply the main force that propels blood through the lungs and the body. The right side is a relatively low pressure system as the blood goes only to the lungs, compared with the left side which supplies the limbs, trunk, and head. This is confirmed by the marked difference in the thickness of the walls of the ventricles—the left is much thicker than the right due to muscular growth (hypertrophy). Overall the adult human heart is usually described as being the size of a fist. It lies in its own lined sac (pericardial cavity) in the space between the lungs known as the mediastinum.

The human (and, more generally, mammalian heart) is an elaboration of the primitive vertebral pattern. The primitive tubular heart received blood at its end towards the tail (caudal) and discharged it from its end towards the head (cephalic end). This tubular heart had five sacculations (subdivisions): sinus venosus, primitive atrium, primitive ventricle, bulbus cordis, and truncus arteriosus. The constriction between the primitive atrium and ventricle later becomes the atrioventricular sulcus in the human heart (Figure 10a).

When the cardiac tube became too long for the pericardial cavity in which it formed an S-shaped loop, its two lower segments (sinus venosus and primitive atrium) and the entering veins came

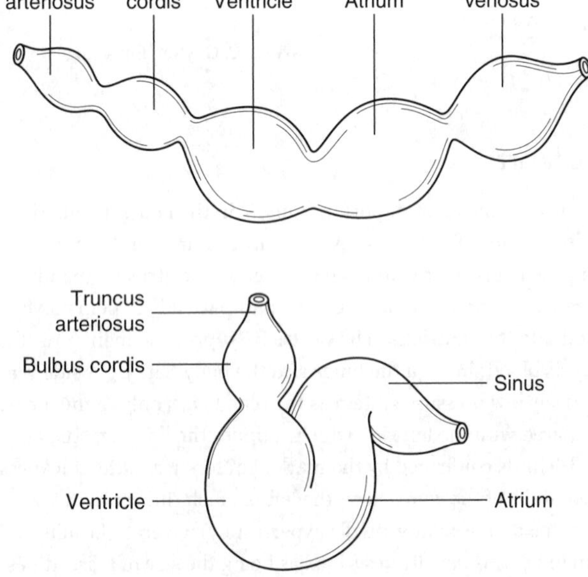

10a. Development of the tubular heart.

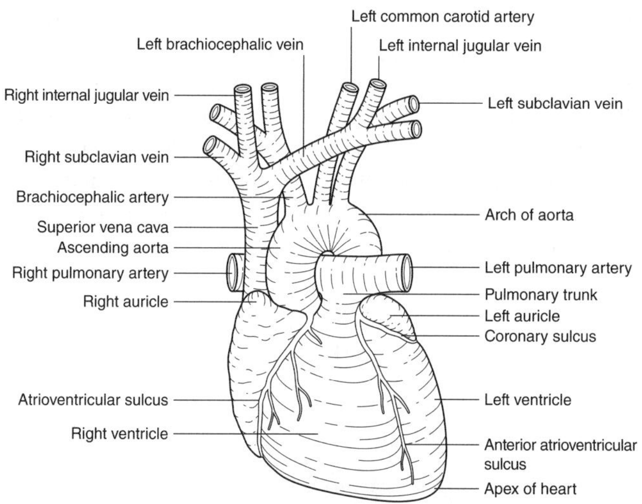

10b. Segments nearest to the tail come to lie behind segments nearest to the head.

to lie behind the three other segments, of which the last (truncus arteriosus) divided to form the ascending aorta and pulmonary trunk. Hence, the atria and entering veins of the adult human heart lie posterior to the ventricles and the emerging arteries. The atria, cramped for space and prevented from expanding forwards, expand laterally on both sides of the truncus (aorta and pulmonary trunk), embracing it (Figure 10b). During evolution the heart also underwent a slight rotation to the left on its long axis. As a consequence of this the right chambers of the heart lie closer to front of the chest wall than the left and a vigorous right ventricle can be sometimes be felt as a 'heave' during clinical examination. The upper end of the right atrium is prolonged to the left of the superior vena cava as an auricular appendage (little ear). Similarly on the left side there is a left auricular appendage that projects from its upper border and passes to the left over the atrioventricular groove.

Blood is collected into the two main veins, the inferior vena cava from the lower limbs and trunk, and the superior vena cava from the head, neck, and upper limbs, to empty in the right atrium (Figure 11). The right atrium pumps blood into the right ventricle and thence to the lungs. Blood returns from the lungs via the four pulmonary veins into the left atrium. From the left atrium it passes through the left ventricle into the aorta and onwards throughout the body.

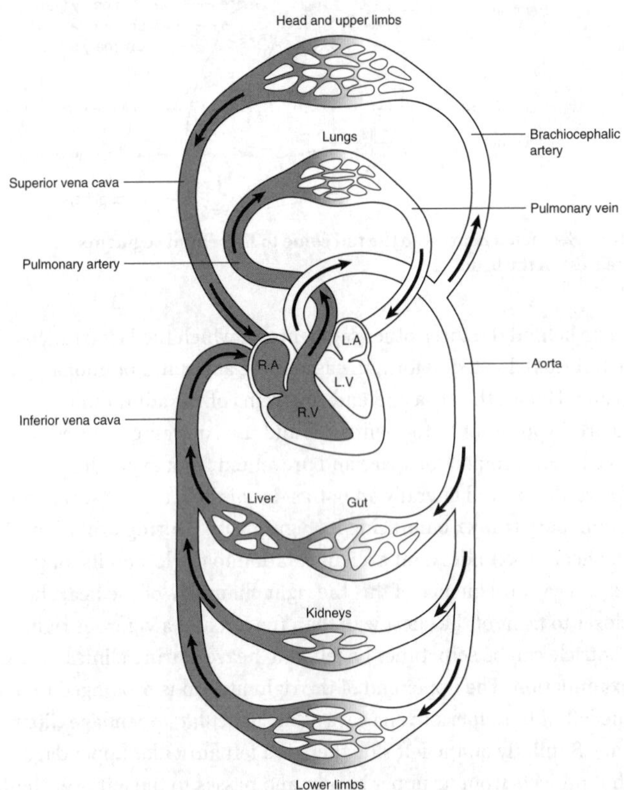

11. A diagrammatic representation of the circulation in man.

The valves of the heart

In order to control the flow of blood throughout the cardiac cycle of systole (contraction) and diastole (relaxation) there are valves between the atria and ventricles and also at the start of the pulmonary trunk and aorta. On the right side the tricuspid valve (three cusps) is between atrium and ventricle and at the exit of the ventricle the valve of the pulmonary artery also has three cusps. On the left side the atrioventricular valve has two cusps only and is called the mitral valve (named by Vesalius because it resembles a bishop's traditional head gear or mitre). The aortic orifice is guarded by the three semilunar cusps of the aortic valve. Between the semilunar cusps and the wall of the ascending aorta are pocket-like sinuses (or dilatations). But for the sinuses, the cusps would stick to the wall of the artery when the valve is open. As blood recoils after ventricular contraction and fills the aortic sinuses, it is forced into the coronary arteries, which supply the heart muscle. Thus the coronary arteries are the first branches of the aorta. The two coronary arteries curve forward one on each side of the pulmonary trunk, sheltered by the corresponding auricle. They occupy the coronary (atrioventricular) and interventricular grooves (sulci), ultimately dividing further to supply the entire heart muscle, although the exact area supplied by each side can vary from person to person. It is these arteries which can be affected by narrowing due to fatty deposits in coronary artery disease—sudden complete blockage leading to the syndrome of myocardial infarction or 'heart attack' (Figure 12). Severe coronary artery disease may be treated by a procedure known as angioplasty (radiologically controlled balloon dilatation of the narrowed segment with the insertion of a stent, a small mesh tube, to keep the vessel open) or coronary artery bypass grafting using the great saphenous vein from the lower limb, or a radial or internal thoracic artery which can be safely transposed for this life-saving procedure.

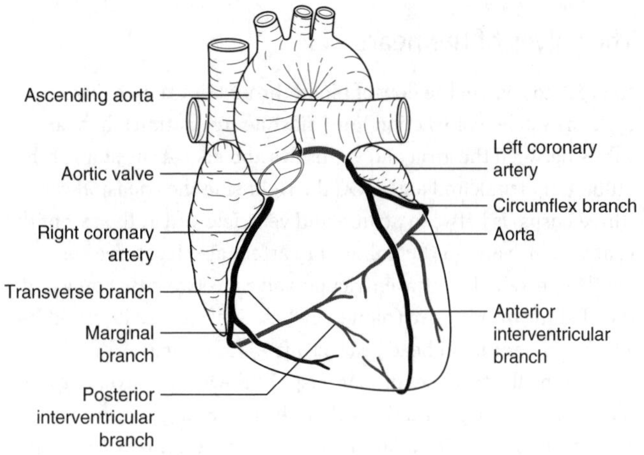

12. The heart showing coronary vessels.

The atrioventricular valves are maintained in the closed position by thin cords (chorda tendineae), which are attached to papillary muscles in the ventricular walls. These muscles contract during systole to stabilize the cusps. Rupture of the chordae tendineae which may occur spontaneously or after myocardial damage causes the valve to leak. This usually affects the mitral valve producing a well-defined systolic 'murmur' (additional rushing sound caused by turbulence of blood flow) heard through a stethescope. Except near the exits the ventricular walls are lined with muscular bundles, or trabeculae carneae ('carnea' is from the Latin for 'flesh').

The heart is endowed with a special system for generating rhythmical electrical impulses to cause regular contraction of the heart muscle and for conducting these impulses throughout the heart (Figure 13). The impulse starts at the sinuatrial node, the pacemaker of the heart, which lies adjacent to the entry of the superior vena cava into the right atrium where auricle meets atrium. The sinuatrial node initiates and regulates the

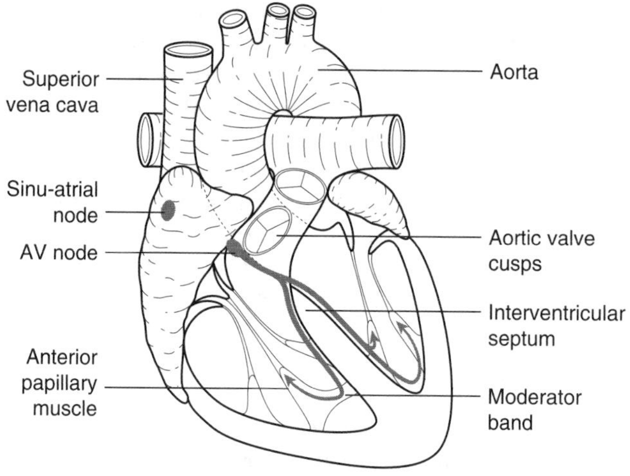

13. The conducting system of the heart.

impulses for the contractions of the heart, giving off an impulse approximately 70 times per minute in most adults. The contraction signal spreads within the heart muscle itself through the atria and converges on the atrioventricular node on the interatrial septum, just above the tricuspid valve. Random propagation of conduction to the ventricles is prevented by a figure-of-eight fibrous insulating skeleton which supports the atria on one side, the ventricles on the other and the valves in between. From the atrioventricular node the main conduction occurs through a single channel—the Bundle of His—which runs through the fibrous skeleton into the interventricular septum, where it divides into left and right bundles to supply each ventricle. These descend to the bases of the papillary muscles. The electrical activity of the conduction system and the heart muscle can be measured by an electrocardiogram (ECG), which measures these currents using electrodes attached across the chest wall and body. Blood pressure is measured using a pneumatic cuff around the right arm and a pressure gauge. When the heart is contracting and is at its maximum, the upper blood pressure reading obtained is the systolic blood pressure, and again when the heart muscle relaxes in diastole the pressure is lowest (diastolic blood pressure). Overall the blood pressure is controlled by a number of factors including the strength of contraction of the heart, the filling pressures in the major veins which return blood to the heart, and the resistance of the arteries. The pressure generated in the heart and also the pulse rate is controlled by the autonomic nervous system (see later, in Chapter 4). This nerve supply to the heart consists of a plexus (compact branching network) of vagal and sympathetic nerves in front and behind the right pulmonary artery which passes to the back of the atria.

When things go wrong

Damage to the conducting system often results from reduced blood flow (ischaemia) and may result in various degrees of heart block with marked slowing of the heart rate. The implantation of a pacemaker—a small electric device which supplies impulses to activate the atria and ventricles—may be necessary to maintain an appropriate ventricular rate of contraction.

A number of by-pass mechanisms in the foetal circulation normally close at birth, but if this process is incomplete it may result in problems for the growing child. Closure of the foramen ovale, which allows blood to pass directly from the right to left atrium, may be imperfect resulting in a 'left to right' shunt. In this condition differential pressures between the two sides of the heart will result in blood flow from left to right. This initially occurs without any cyanosis (blue or purple discoloration of the skin due to lack of oxygen in the tissues) as the high pressure blood that is moving from the aorta is well-oxygenated. 'Right to left' shunts, where de-oxygenated blood is moving into the aorta under high pressure, lead to cyanosis. A persistent ductus arteriosus (which connects the left pulmonary artery to the aortic arch) will allow blood flow from the left to right side meaning that blood circulates through the heart too often. This results in raised pulmonary blood pressure. There is no initial cyanosis, but strain is placed on the right side of the heart with adverse consequences. Treatment is closure by surgical ligation.

As blood tends to carry the sound in the direction of flow, the areas for listening to (auscultation) the heart valves are superficial to the chamber or vessel into which the blood has passed, and they are generally in line with the valve orifice. Thus those at the right side of the heart are listened for around the sternum, while those on the left are listened for over towards the left axilla. If the valves become narrowed (stenosed) or leaks (incompetent) murmurs will be heard during systole or diastole, depending on the valve affected. If these vibrations are very large, these leaks can not only be heard, but also felt as a 'thrill' on the front of the chest.

Most people with progressive constriction of their coronary arteries, will experience cardiac pain, called angina pectoris, whenever the load on the heart becomes too great in relation to coronary blood flow. This pain is usually felt beneath the upper sternum and is often also transferred (referred) to other regions of the body such as the left arm and left shoulder, but also frequently to the neck and even the side of the face. The reason for this distribution of pain is that the heart develops in the neck in the embryo as do the arms. Therefore both of these structures receive pain fibres from the same spinal cord segments.

Respiration and the lungs

As with the beating of the heart, breathing is synonymous with human life. The body needs a constant supply of oxygen. The lungs, surrounded by their pleural cavities, occupy the chest on either side of the heart (Figure 14). As with all cavities in the body there are two layers of lining, with the inner (visceral) pleura, closely following the contours of the lungs; and the outer (parietal) pleura, lining the ribs and covering the diaphragm. The diaphragm is the muscular partition between the thorax and abdomen. It is attached peripherally to the upper six ribs and

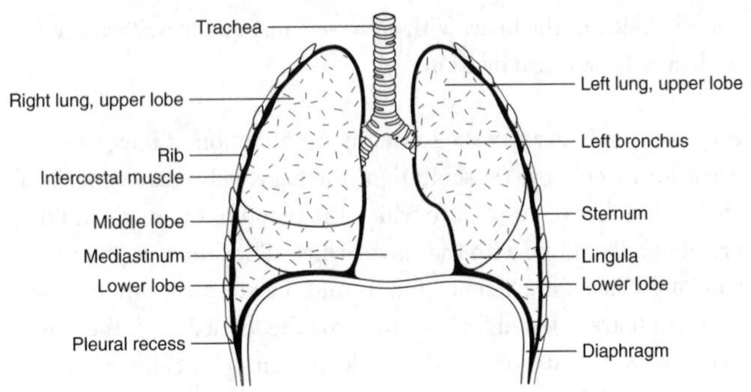

14. The lungs and the diaphragm.

xiphisternum (cartilaginous extension of the sternum) and has two vertical attachments (the crura) to the upper lumbar vertebrae. There is a central tendon, which is partially fused with the undersurface of the pericardium. The diaphragm receives its entire motor nerve supply from the phrenic nerve (cervical nerves 3, 4, and 5) whose long course from the neck follows the embryonic migration of the muscle from the cervical region. The diaphragm acts like a pair of bellows that suck air into the lungs and then expel it. This action is accompanied by 'pump-handle' movements of the upper ribs which are attached to the sternum and 'bucket-handle' movements of the lower ribs. Expiration (breathing out) is largely a matter of elastic recoil. Typically, each intercostal space (those between the ribs) contains three muscles and a neurovascular bundle (i.e. intercostal nerve, artery, and vein). Contraction of the intercostal muscles prevents both indrawing of the intercostal spaces during inspiration and bulging of the spaces during expiration.

The trachea, or wind pipe, is an elastic tube which starts below the larynx and is easily felt in the neck. It divides into right and left bronchi at the level of the sternal angle. The trachea is kept open by about twenty U-shaped rings of hyaline cartilage, which are incomplete at the back. The bronchi have the same structure as the trachea. They descend towards the base of the lung giving off branches, which in turn branch and rebranch like those in a tree. As they descend, they lose their hyaline support, becoming bronchioles, and finally terminate as thin-walled alveolar air sacs (alveoli) where oxygen is absorbed.

The right and left lungs are sponges composed of elastic tissue. Branches of the bronchus, pulmonary artery, and vein divide to form a network within this highly elastic framework. The right lung is divided by two complete fissures into three separate lobes; the left is divided by one fissure into two lobes. The left lung is partially deficient in front where it overlies the heart and pericardium. This deficiency is known as the cardiac 'notch'. The site of entry of the main vessels and bronchi into each lung is called the hilum, which may contain lymph nodes, filtering the lung's contents. The lung tissue has a dual blood supply—the main blood flow is of de-oxygenated blood through the pulmonary arteries from the right side of the heart. To supply oxygen to the tissues, they are provided with arterial blood by bronchial arteries which arise either from the aorta or an intercostal artery and run with the bronchi.

The lymphatic system

The lymphatic system plays an important role in the circulation by draining fluids that accumulate in the extracellular space during capillary exchange and returning them to the blood. However, it is much more than a drainage network. It helps in defence against bacteria and cancerous tumours as well as transporting fats. The vessels are very thin, and even the main lymphatic vessel of the body, the thoracic duct, is only comparable to a small vein in size. Although lymphatic vessels were well-known to the ancient Alexandrian anatomists in the 3rd century BC, it was not until 2,000 years later that they were 'rediscovered' and described in some detail. In 1627 Gasparo Aselli (1581–1626), Professor of Anatomy at Pavia, a city near Milan, noted them in the mesentery (a membranous fold joining the intestine to the posterior abdominal wall) of a well-nourished dog. His attention was drawn to them as a large proportion of the fat is absorbed in the small intestine so that after a meal the mesenteric vessels become filled with a milky fluid called 'chyle' which makes them visible to the naked eye. The term 'lymphatic' was first used by the Danish physician, Thomas Bartholin (1616–80), in 1653.

Most of the constituents of the blood plasma (noncellular part of blood) pass through the walls of the capillaries into the tissue spaces for the nourishment of the tissues and, after undergoing metabolic change, they pass back again and return to the heart via the veins. Some fluid passes into lymphatic capillaries where it is drained by lymph vessels through lymph nodes to the great veins at the root of the neck, finally rejoining the blood stream. Lymphatic fluid is clear and colourless. Lymphocytes are added to it as it passes through the lymph nodes. Lymph capillaries occur only where there are blood capillaries and form a closed network. They tend to follow veins but are more numerous. This intimate relationship between lymphatic and venous channels is to be expected in view of the embryonic derivation of lymphatic vessels from the venous system. The walls of lymphatic vessels are similar to those of veins and like veins contain one-way valves. Where lymphatics are present they usually run in the looser areas of connective tissue. Hence their distribution is determined to a large extent by fascial planes. Lymph nodes (lymph glands), which are found throughout the body, vary in size from that of a pin's head to that of an olive and are somewhat flattened in shape. They are pink in living humans, although those draining the lung may become black from inhaled carbon and, after a meal, those draining the intestines are white from emulsified fat. The nodes act as filters for lymph and factories for lymphocytes. Lymphatic nodules similar to those found in a lymph node are also present in a number of other structures that can loosely be described as 'lymphoid tissues'. These are the spleen, thymus, tonsils, and distal small intestine where the nodules are known as Peyer's patches.

The flow of lymph in an immobile limb is almost negligible, but during muscular activity it becomes very active. Lymph flows in the same direction as venous blood, towards the heart. The two largest lymphatic vessels, the thoracic duct on the left and the right lymph duct drain into the junction of the subclavian and internal jugular veins on each side of the neck. The thoracic duct drains a much more extensive area of the body and is therefore much larger than the right lymph duct. Vesalius called the thoracic duct 'vena alba thoracis', the white vein of the thorax. In the past, these very slender lymphatic vessels and nodes could only be seen by the injection of dyes and contrast media (lymphangiography). However, this method has been largely superseded by modern imaging techniques such as CT (computerized tomography) and PET (positron emission tomography) scans.

The breasts

The breasts (mammary glands) are the most prominent superficial structures in the anterior thoracic wall. They are rudimentary and functionless in men. Human breasts are unusual because of the presence of significant amounts of adipose tissue around the glandular tissue. This fat creates the conspicuous shape of the breast. Although body fat is the source of much of the energy used to produce milk it is not apparent why humans have any more reason to place the fat next to the glands than do other species. In view of the erotic visual value of breasts to males in many societies it has generally been argued that their form is the product of sexual selection for the attraction and retention of male attention.

Breasts contain modified sweat glands called mammary glands which produce milk at childbirth. A breast contains 15–20 lobes of compound areolar glands, each lobe resembling a bundle of grapes on a long stalk. The cells of the glands secrete milk, which flows along lactiferous ducts that converge towards the nipple. The breasts are firmly attached to the dermis of the overlying skin by substantial skin ligaments, the suspensory ligaments of Cooper

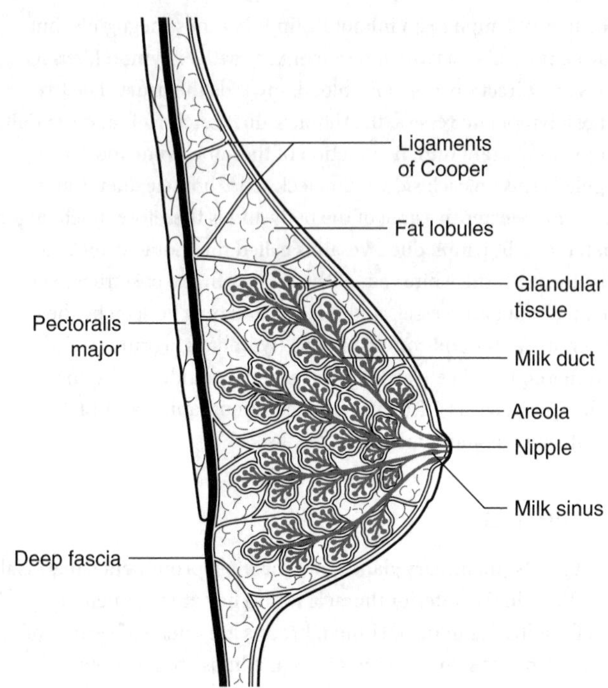

15. Adult female breast.

(Figure 15). Extra nipples are occasionally found in humans, and they typically arise along the same 'milk lines' that characterize multi-teat species such as dogs and pigs. At an early embryonic stage the genitalia are the same in boys and girls. Male nipples are thus a genetic remnant that does not develop after the male hormone testosterone starts to function at about sixty days of embryological development.

When things go wrong

Pneumothorax is the presence of air between the visceral and parietal layers of the pleura, which are normally separated by a thin film of fluid. A common cause is a fractured rib or penetrating injury, but they may also occur spontaneously. In hospital practice it may occur as a result of the insertion of needles for placing lines into the large veins such as the sublavian, for monitoring venous pressure, intravenous feeding, or implantation of a cardiac pacemaker. Air passes in and out of the chest wall with each breath but no effective ventilation occurs. When air in the pleura is under pressure the situation is called a 'tension' pneumothorax and is a medical emergency. The latter is produced by the valve-like mechanism, which is inherent in the structure of the lung. Positive pressure generated within the airways forces air out into the pleura, which cannot return through the collapsible peripheral alveoli. Blood can also collect in the pleural space by itself or with air to result in a haemothorax or haemopneumothorax, while large collections of pus can collect in the setting of lung infection—known as an empyema. Large symptomatic pneumothoraces require the insertion of a chest drain. Curiously, the buffalo has only a single pleural space. This fact was advantageous to native Americans who knew that a single arrow in a buffalo's chest would ultimately lead to the animal's death.

In a manner similar to the pleura, fluid may collect in the pericardial cavity between the fibrous and parietal layers from inflammation or injury (pericardial effusion). This external pressure may interfere with the normal function of the heart and is known as cardiac tamponade. Emergency treatment requires aspiration of the pericardial space.

Acute lymphangitis occurs when an infection, often streptococcal, spreads from a local site such as a toe to the regional lymph nodes draining that area. It results in red blushes and streaks in the skin, corresponding to the inflamed lymphatics.

Congenital deficiency of lymphatics may present at variable stages of life, at birth, puberty, or in adult life. There is an accumulation of extracellular, extravascular fluid, mainly in the subcutaneous tissues, which is called lymphedoema. In the advanced state it characteristically does not indent on pressure (nonpitting oedema). It often affects a foot and ankle only, but may be more extensive and affect a whole limb. Lymphoedema of the whole upper limb used to be a feature after radical mastectomy for breast carcinoma with clearance of axillary lymph nodes.

Chapter 4
Communication and control

It would be idle to think that the study of the structure and function of the central nervous system is not difficult.

David Bowsher, Neuroanatomist, Liverpool

All vertebrates, despite their many outward differences, have a similar basic body plan—the segmented backbone or vertebral column surrounding the spinal cord with the brain at the head end enclosed in a bony or cartilaginous skull. The human nervous system contained within this skeleton is the most highly developed in the animal kingdom. It consists of two components that are anatomically separate but closely related in function: the *central nervous system* and the *peripheral nervous system*. The central nervous system is composed of two cerebral hemispheres, the brainstem and the spinal cord. The peripheral nervous system may be further subdivided into a *somatic* division and an *autonomic* division. The somatic division includes the motor supply of the skeletal muscles and the sensory innervation of the skin, the muscles, and the joints. This division of the peripheral nervous system mediates voluntary movement and much of perception. The autonomic system deals with activities that are primarily automatic such as control of heart rate and blood pressure, which are involuntary. Our precious and delicate nervous system is protected by the skull and vertebral column. The brain lies comfortably within the skull cavity (cranium), cushioned by a series of membranes and cerebrospinal fluid, which circulates within the brain and subarachnoid space (see later). The brain has the monopoly on the cranial cavity, to such an extent that not even fat is allowed within it.

The skull

The skull can be divided for descriptive purposes into a facial and cranial part. The cranial bones form the walls of the cavity that contains the brain; and the facial bones are situated below the front portion of this 'brainbox' (Figure 16a). At birth, the face is considerably smaller in proportion to the skull than in the adult as the teeth have not erupted and the nasal sinuses are still undeveloped.

The shape of the inside of the skull mirrors that of the brain. There are three cranial hollows (fossae), anterior, middle, and posterior with a large opening in the posterior fossa (foramen magnum) for the exit of the spinal cord. In humans, because we are upright, the foramen magnum is near the centre of the posterior cranial fossa, but in apes and all other quadrupeds it lies near the posterior end of the skull. The three fossae descend

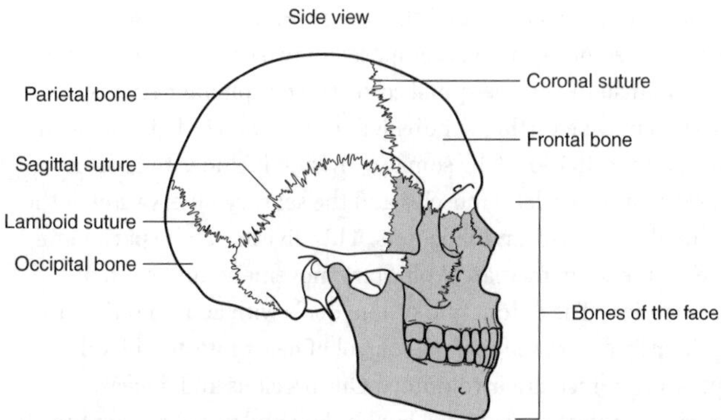

16a. Skull showing facial bones and 'brainbox'.

backwards like the steps of a staircase. The middle cranial fossa consists of a deep concavity on each side which are joined across the waist by the sella turcica (Turkish saddle) for protection of our most important endocrine gland, the pituitary, which orchestrates the other endocrine glands. The bone of the skull is relatively elastic in consistency: thus a blow may transmit force to and injure the underlying brain without fracturing bone. The interior of the cranial fossae is perforated by various foramina (holes) and fissures to allow for the passage of vessels and pathways for nerves (Figure 16b).

The skullcap, calvaria, or the bony area beneath the hairy part of the scalp has inner and outer tables with cancellous or spongy bone sandwiched between them that contains red bone marrow. This is one of the sites of persistent red marrow in the adult skeleton. The skull cap is simply constructed from four bones: the frontal bone in front, the occipital bone behind, and

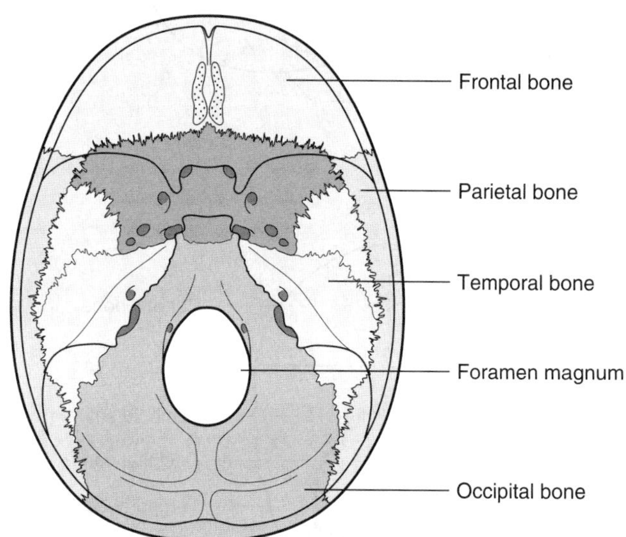

16b. Interior of skull.

the two parietal bones in between, one on each side. The *sagittal suture* (joint between two bones) is in the midline between the parietal bones, and the occipital bone joins the parietal bones at the *lambdoid suture*. The *coronal suture* divides the frontal from the parietal bones (Figure 17). A knowledge of the suture lines is

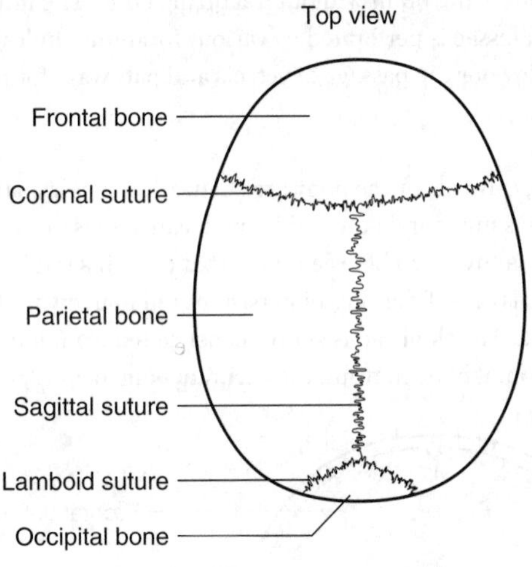

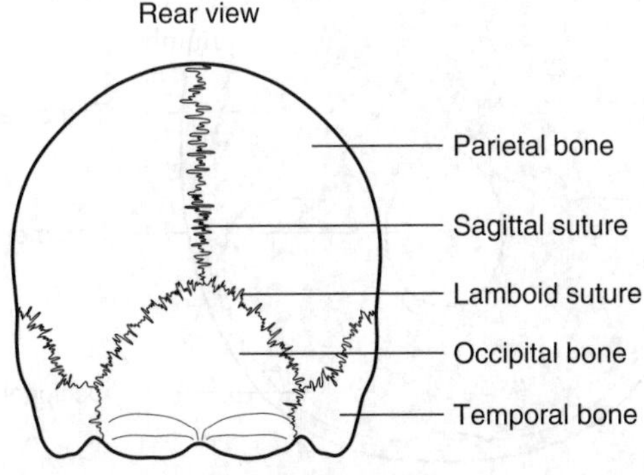

17. The skull from the back and on the dome.

important in interpreting skull X-rays, when they may otherwise be mistaken for fracture lines. In the neonate the bones of the skull are mobile on one another. They do not interdigitate in sutures as in the adult, but are separated by linear attachments of fibrous tissue and at their corners by larger areas known as *fontanelles*—clear soft tissue defects in the bony skull. This arrangement allows for moulding of the skull during birth. The anterior and posterior fontanelles are readily seen as they lie in the midline of the vault. The anterior fontanelle usually closes by eighteen months and the posterior one by the end of the first year.

The brain

The brain develops as three swellings at the anterior end of the neural tube on the dorsal surface of the embryo (the longer posterior portion of the neural tube gives rise to the spinal cord). These primary vesicles develop into the forebrain, midbrain, and hindbrain. The cerebral hemispheres develop from the forebrain. In the adult these dwarf the remaining portions of the brain and occupy a very large part of the cranial cavity. The hemispheres are separated by a deep longitudinal fissure and in the depths of this they are joined by a thick band of communicating nerve fibres called the corpus callosum—the communication bridge between the cerebral hemispheres (Figure 18). The convoluted surface of the human cortex can be divided into a number of functional regions but the most basic and simple division is between areas serving motor and sensory functions. Sensory cortical areas are defined by the type of information they receive. Different areas are specialized to receive and process information coming from particular sensory organs and structures. Information from the eyes passes to the most posterior part of the occipital lobe as the primary visual cortex (lobes are named after the bones of the skull which overlie them). Information from the body and skin called 'somatosensory' information passes to a strip of parietal lobe known as the somatosensory cortex. Movements of the body are controlled by the primary motor cortex. The motor areas occupy

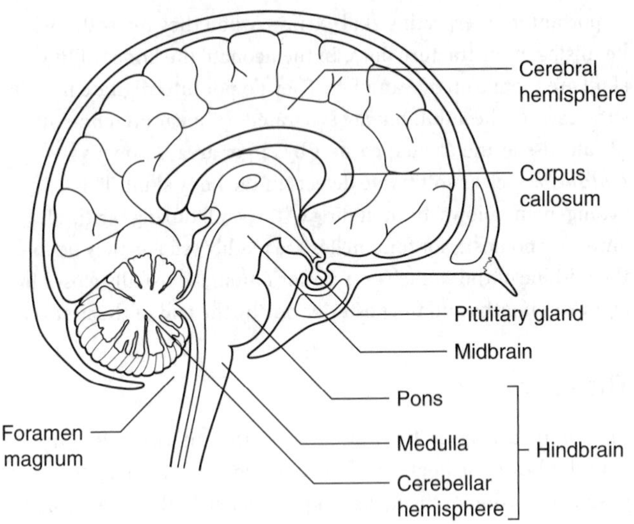

18. Sagittal section of brain.

the strips of cortical surface just in front of the prominent central sulcus which is located roughly halfway between the occipital pole and the anterior part of the frontal cortex. It is the boundary between the frontal and parietal lobes. The sensory cortex lies behind this central sulcus. A distorted image of the human form known as Penfield's homunculus reflects the differential allocation of cortical surface to sensory and motor functions of different parts of the body (Figure 19). For example, because of the density of nerve fibres supplying information from the fingertips, these are hugely over-represented in the brain compared to their actual size.

As shown most starkly by human stroke victims the left side of our brain controls the right side of our body and vice versa. The diagonal control is mediated by a crossing of nerves from the motor cortex in the brain stem. This arrangement serves no obvious function either anatomically or physiologically—but it

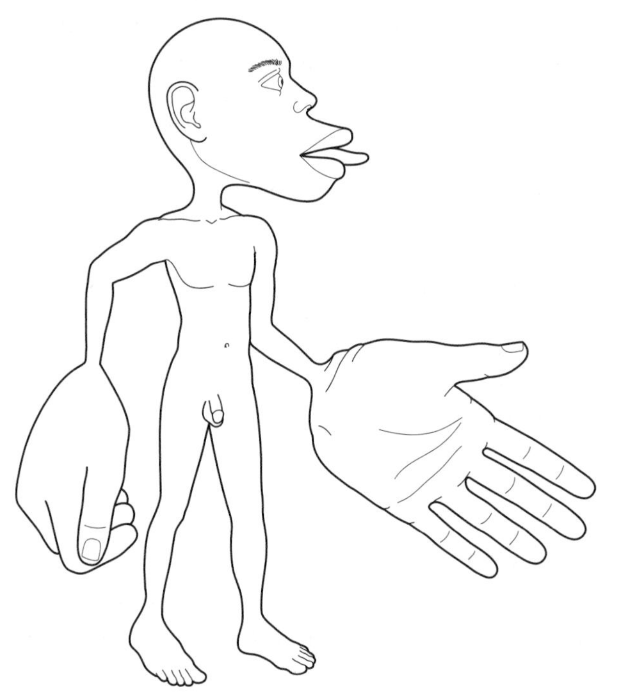

19. Wilder Penfield's homunculus.

may have proven handy long ago in our prefish ancestors. One idea is that crisscrossing arose as part of a 'coiling reflex'. Possible danger detected on one side of the body, such as a predator's shadow, would have instantly caused the muscles on the opposite side to contract thereby turning the body from potential harm. Another theory holds that diagonal wiring evolved in conjunction with undulatory locomotion to coordinate sinusoidal swimming.

The midbrain is short in extent. It is characterized anteriorly by two stout pillars leading from the forebrain to the hindbrain, called cerebral 'peduncles'. The hindbrain which consists of the pons and medulla with the cerebellar hemispheres joined to their posterior aspect, lies in the posterior cranial fossa. The medulla is continuous with the spinal cord at the foramen magnum. These parts of the brain are responsible for co-ordination of various critical functions such as balance, breathing, and blood pressure.

The cavity of the neural tube is retained to adulthood and represented within the brain structure as the ventricular system, with the two largest ventricles obvious in the hemispheres in images of the brain by CT (computerized tomography) or magnetic resonance imaging (MRI) scan. Cerebrospinal fluid (CSF) circulates in this system but only communicates with the subarachnoid space between pia and archnoid at the fourth ventricle through three openings. The central canal of the spinal cord (also part of the system) normally does not remain patent beyond the early postnatal period.

The meninges

The meninges consist of three layers of membranes which envelop the brain and spinal cord. Their names indicate their properties. The dura mater (hard mother) is tough, the arachnoid mater is like a spider's web, and the pia mater (tender mother) clings faithfully to the brain and cord surfaces like a skin following all the irregularities. Between the dura mater and arachnoid mater there is a potential space, the subdural space. As mentioned above, between the arachnoid and pia there is an actual space, the subarachnoid space filled with CSF. The arachnoid is attached to the pia by loose, scattered fibrous threads. The cerebrospinal fluid is produced by a modified vascular structure called the choroid plexus, which is present in each of the ventricles. It is eventually absorbed through specialized structures (arachnoid villi) along the dorsal midline of the forebrain and returned to the venous circulation. In a similar fashion the cerebrospinal fluid in the spine is reabsorbed through thin walled venous sinuses between layers of dura mater.

The CSF has three functions: it acts as a shock absorber, it supplies nutrients to the tissues of the nervous system, and it removes waste products. Clinically, access to the CSF for diagnostic purposes can be obtained by lumbar puncture, which is by passing a fine needle into the subarachnoid space at the spinal level of the interspace between lumbar vertebra 3 and 4, or between 4 and 5. The spine must be fully flexed (with the patient on his or her side) so that the vertebral interspaces are opened to their maximum extent. The needle is placed in the midline at right angles to the spine. It then traverses the supraspinous and interspinous ligaments and finally with a definite 'give' pierces the dura. If normal, the CSF appears as clear liquid. In disease, it may be bloody or contain pus.

The inner layer of the dura is reduplicated to form four inwardly projecting folds which partially subdivide the cranial cavity into compartments, and being taut they prevent shifting of the brain. Two of the folds, the falx cerebri and falx cerebelli are sickle shaped and lie in the midline. The tentorium cerebelli, shaped like a bell tent with open flaps, forms a roof for the cerebellum and a floor for the posterior parts of the cerebrum. Between the layers of the dura lie venous sinuses, filled with blood from tributaries from the cortical veins of the brain and veins from the skull. They groove the internal surface of the skull. All the sinuses drain into the internal jugular veins on each side.

Blood supply of the brain and spinal cord

Although the brain only accounts for only 2 per cent of the body weight, it receives 15 per cent of the total cardiac output. Nerve cells can only live a few minutes without oxygen. In contrast a chimpanzee's brain makes up less than 1 per cent of its body weight and is supplied by a somewhat smaller fraction of its blood supply, probably 7–9 per cent. The entire blood supply of the brain and spinal cord depends on two sets of branches from the main arterial trunk of the body, the dorsal aorta and its arch. The vertebral arteries arise from the subclavian arteries; the internal carotid arteries are branches of the common carotid arteries from the arch of the aorta. The name 'carotid' is derived from the Greek word for stupor because it was known that bilateral compression of these vessels in animals or man would cause unconsciousness. The anterior and posterior spinal arteries descend in the pia from the intracranial part of the vertebral arteries. They are reinforced at each vertebral level by branches from the cervical parts of the vertebral arteries, the posterior intercostal and lumbar arteries. Thus the spinal cord is supplied segmentally.

Both the internal carotid and vertebral arteries enter the skull at its base. Anterior to the spinal cord and brainstem, the internal carotid arteries branch to form two major cerebral arteries: the anterior and middle cerebral arteries. Posteriorly the right and left vertebral arteries join on the front surface of the pons to form the midline basilar artery. The basilar artery joins the blood supply from the internal carotids in an arterial ring at base of the brain at the level of the pons, called the Circle of Willis (Figure 20). The posterior cerebral arteries arise at this confluence as do two small bridging arteries, the anterior and posterior communicating arteries. Because of the connection between the two major sources of cerebral blood supply, the chances of any region of the brain continuing to receive blood if one of the major arteries is occluded is improved and protected against the possibility of a stroke. There is considerable anatomical variation in the Circle of Willis. Based on a study of 1413 brains the classic anatomy of the circle is only seen in one-third of cases. Thomas Willis (1621–75), an English doctor and Professor of Natural Philosophy at Oxford University, was the first person to illustrate the circle named after him. The pictures in his book *Cerebri Anatome* (1664) were drawn by Christopher Wren (now best known for his later work as architect of St Paul's Cathedral), who was a skilful artist and, at this stage, a pure scientist (by the standards of the time), focusing on astronomy, physics, and anatomy.

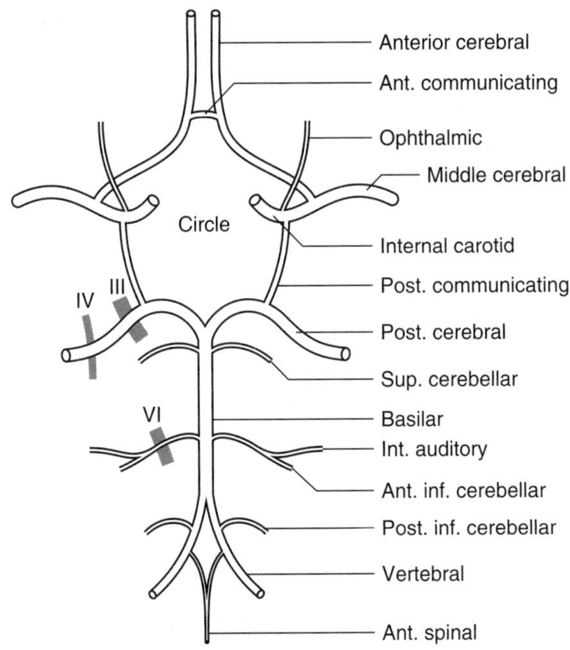

20. The Circle of Willis, which lies on the undersurface of the brain around the pituitary fossa.

Cranial nerves

The brainstem is the source of the cranial nerves that deal with sensory and motor functions in the head and neck. There are twelve pairs. Thomas Willis was the first to number the cranial nerves in the order they are now usually enumerated by anatomists. They are usually designated by Roman numerals. All exit from the ventral (undersurface) of the brain except the fourth, the trochlear nerve, which supplies the tiny superior oblique muscle that is attached to the eyeball. The first two nerves, the olfactory (I) and optic (II) are extensions of the brain and surrounded by dura and arachnoid. They are sensory nerves that supply, respectively, smell signals from the nose and visual information from the eyes.

The third, fourth, and sixth nerves are highly co-ordinated and responsible for supplying muscles which move the eyes. In addition the third nerve has fibres that control the size of the pupil (contracting in response to bright light) and tension on the lens ('accommodation' for focusing on near objects). The fifth nerve, the trigeminal, has three divisions—ophthalmic, maxillary, and mandibular—each primarily sensory for the face, including the eye and surface of the head. There is also a motor root, which accompanies the lowest, mandibular, division to supply the muscles around the jaw used for chewing. The seventh nerve, the facial, is responsible for all the complex movements of the face, together with salivation and production of tears. In its most delicate role, it supplies a tiny muscle in the ear, the stapedius, which controls the vibration of the tiny ossicles that transmit sound. It is also a sensory nerve, supplying the tongue and largely responsible for the sense of taste. The eighth nerve, the vestibulocochlear, receives signals from the inner ear and vestibular apparatus, regulating balance and hearing. The ninth nerve, glossopharyngeal, is involved with taste, salivation, and swallowing. The tenth nerve, the vagus or 'wandering nerve', is a long nerve which has many roles in diverse parts of the body. It acts in the throat in swallowing and movements of the soft palate, in the larynx in production of the voice, continues down through the thorax to provide essential communication to and from the heart and lungs, and goes on to supply the abdominal viscera. The eleventh nerve, the accessory, supplies muscles which turn the head (sternomastoid muscle) and move the shoulders (trapezius muscle). Unlike all the other nerves, which arise entirely from the brain itself, this contains an additional root from the upper spinal cord, which ascends up into the skull to merge and create the full nerve. The twelfth nerve, the hypoglossal, moves the tongue. Nerves III, VII, IX, and X contain fibres that are part of the 'parasympathetic' component of the autonomic nervous system, as well as motor and sensory fibres. A detailed knowledge of the cranial nerves is part of the armamentarium of all good clinicians as these nerves deal with sensory and motor function in the head and neck. Understanding the internal anatomy of the brainstem, although complex, is generally regarded as essential for neurological diagnosis and the practice of clinical medicine.

The spinal cord

The spinal cord lies within the vertebral canal and is about the diameter of a finger and the length of a femur. It has the same meningeal coverings as the brain and is bathed in the same cerebrospinal fluid. In the foetus the spinal cord extends down to the coccyx, but as development proceeds, owing to the greater growth of the vertebral column, it is drawn upwards, so that at birth it extends only to the level of the third lumbar vertebra and in the adult to the upper part of the second lumbar vertebra. A strong glistening thread, the filum terminale, largely composed of pia mater, attaches the end of the cord to the back of the coccyx, even in the adult. A broad band of pia, the ligamentum denticulatum projects from each side of the cord and lies between the ventral and dorsal nerve roots. From the free margin of this ligament strong tooth-like processes, one for each segment, pass through the arachnoid to become firmly attached to the dura. The cord is elastic due to the pia attached to it and stretches with flexion of the vertebral column.

The spinal cord begins just inferior to the foramen magnum of the skull, where it is continuous above with the medulla of the hind brain. It ends opposite the second lumbar vertebra. In shape it is a somewhat flattened cylinder and shows two gentle swellings along its course. The upper of these, the cervical enlargement, is related to the nerves supplying the upper limbs (brachial plexus). The lower is called the lumbosacral enlargement and is related to the nerves supplying the lower limbs (lumbosacral plexus). Below this the cord tapers to a point called the conus medullaris. The nerve roots lying below the termination of the spinal cord form collectively the *cauda equina* (the leash of nerves resembles a horse's tail) occupying the lower lumbar and sacral portions of the

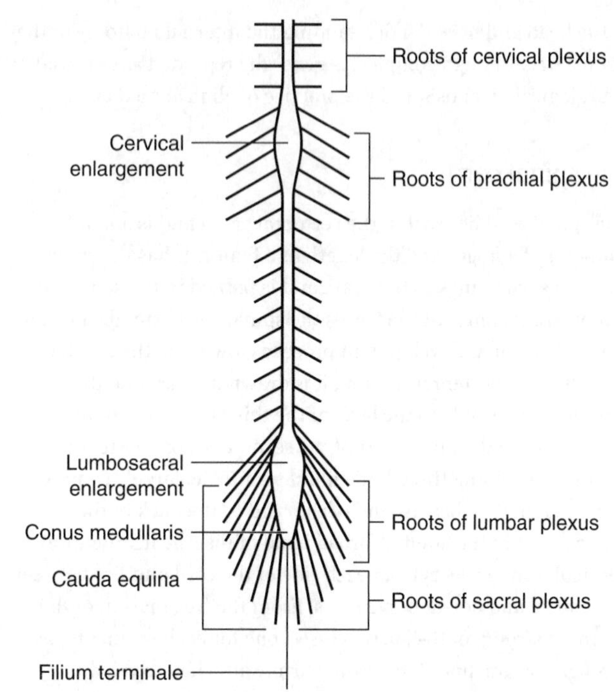

21. The parts of the spinal cord.

vertebral canal (Figure 21). The cord has, in cross-section, an H-shaped field of grey matter (which contains nerve cell bodies), enveloped in a zone of white matter. The anterior and posterior limbs of the H, called the anterior and posterior horns or grey columns, divide the white matter of each side into anterior, lateral, and posterior white columns. They are white because the nerve fibres are covered with a myelin sheath which is an insulating material that greatly increases the speed at which electrical signals are transmitted, a crucial feature because function in the nervous system depends on information being transmitted at high speed. A groove in front and a septum behind separate the right and left sides of white matter. A canal, the central canal of the cord, continuous with the ventricles of the brain, runs through the grey matter in the newborn.

At each vertebral level, an anterior row of rootlets converge laterally from the cord to form the anterior nerve roots; similarly the rootlets of the posterior row unite to form the posterior nerve root. The anterior (ventral) and posterior (dorsal) roots pierce the dura about 2 millimetres (mm) apart. Each dorsal root is swollen because it contains the bodies of sensory nerve cells; the swelling is called a spinal root ganglion. Just beyond this, the two roots unite and their fibres mingle to form a mixed peripheral nerve. This in turn soon divides into a large ventral and a small dorsal ramus (branch) for the ventral and dorsal trunk. The length of cord to which the rootlets of one pair of spinal nerves is attached is called a spinal segment. Since there are thirty-one pairs of spinal nerves the cord has thirty-one segments (contrast twelve pairs of cranial nerves). At each segment, the dorsal nerve root, which transmits impulses to the cord, is known as afferent or sensory. A ventral nerve root, which transmits impulses from the cord, is efferent or motor. Afferent or sensory spinal nerve fibres bring impulses from nerve endings in skin, muscles, joints, etc., to the spinal ganglia and thence via the rootlets to the cord. The efferent nerves end in motor end plates of voluntary muscle to produce muscular contraction.

The autonomic nervous system

Our knowledge of the autonomic nervous system was only acquired in the late 19th century mainly as a result of work by two British physiologists, Walter Gaskell (1847–1914) and John Langley (1852–1925) at Cambridge University. This is an involuntary (hence the name autonomic) motor system which regulates the functions of internal organs and has two components—sympathetic and parasympathetic. The sympathetic system as a whole is a catabolic system expending energy as in the 'fight or flight' response to danger by increasing the heart rate and contractility, and shunting blood to the muscles and heart. The parasympathetic system is anabolic ('rest and digest'), conserving energy by slowing the heart rate and in promoting the digestion and absorption of food. The cell bodies of preganglionic sympathetic fibres lie in the lateral horn of the spinal cord midway between anterior and posterior horns at spinal cord levels of Thoracic 1 to Lumbar 2. The axons from these spinal preganglionic neurons (nerve cells) typically extend only a short distance and end in a series of connecting stations known as paravertebral sympathetic ganglia arranged in a chain that extends most of the length of the vertebral column. The nerves of the parasympathetic system occupy comparable positions at spinal cord levels Sacral 2–Sacral 4. In addition, as mentioned earlier, some cranial nerves have parasympathetic components (Figure 22). The parasympathetic ganglia, innervated by outflow from both cranial and sacral levels are in or near the end organs they serve. In this way they differ from the ganglionic targets of the sympathetic system, where the paravertebral chain and prevertebral ganglia are located relatively far from their target organs.

When things go wrong

When a penetrating injury occurs to the skull, the mechanism for the underlying damage to the brain can easily be understood. When a high-velocity acceleration/deceleration or shearing injury occurs, the effects of the primary injury may be much more diffuse and remote from the point of impact. The brain in its normal physiological state is afforded limited movement within the cranium, movement which is cushioned by the cerebrospinal fluid. When impact occurs and a decelerative force is applied to the cranium, considerable energy is transferred to the brain as it collides with the internal surfaces of the skull. A *contra-coup* injury occurs when the brain is injured on the opposite side to the site of the blow to the skull. Trauma may also be accompanied by bleeding from small or larger blood vessels. Haemorrhage may occur from rupture of the middle meningeal artery resulting in an 'extradural' haematoma, or tearing of fragile venous sinuses in the

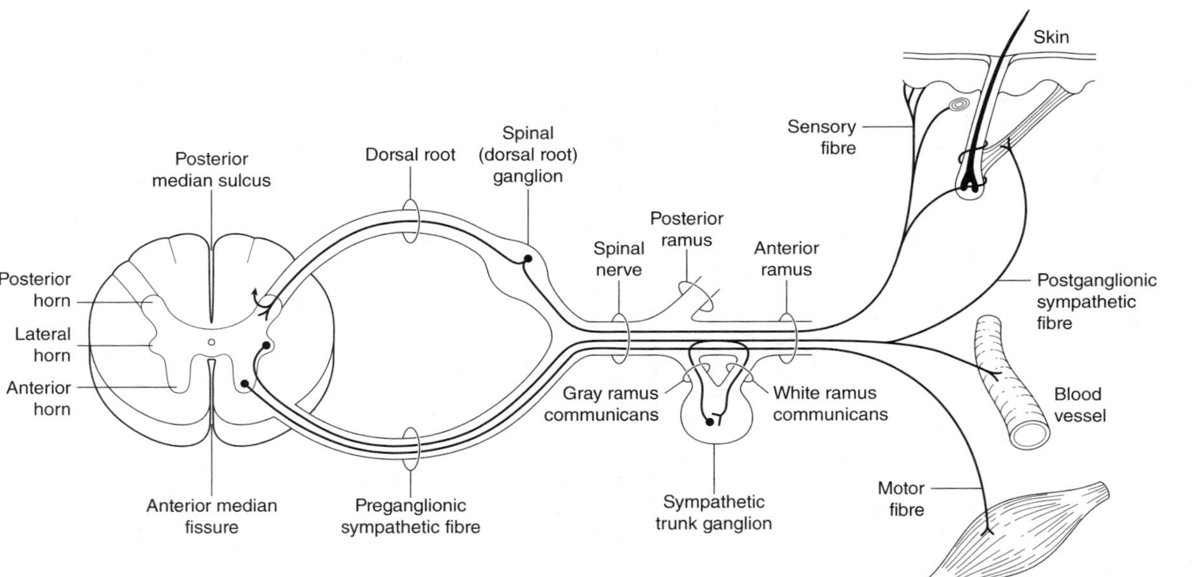

22. The spinal cord and connections to the autonomic nervous system.

elderly giving a 'subdural' haematoma. In both cases, pressure due to the bleed can give rise to acute or more gradual impairment of neurological function.

Stroke is the most common neurological cause for admission to hospital in the elderly. The term refers to the sudden appearance of a limited neurological deficit, such as weakness or paralysis of a limb, or the sudden inability to speak. The onset of the deficit within seconds, minutes, or hours marks the problem as a vascular one. There are three main types: thrombotic, embolic, and haemorrhagic. A thrombotic stroke is caused by local clot formation. A reduction of blood flow may occur when an embolus (an object loose in the blood stream) dislodges from the heart or the inside of an artery and travels to a cerebral artery where it forms a plug and thus causes an embolic stroke. A haemorrhagic stroke occurs when a cerebral blood vessel ruptures as a result of high blood pressure, a congenital aneurysm (bulging of a vessel), or injury. The exact form of the deficit and the capacity for recovery depends on the territory supplied by the vessel and the pathways crossing that region.

Hydrocephalus occurs due to the accumulation of cerebrospinal fluid within the skull. It may be due to excessive production of fluid, obstruction in some part of the route along which the fluid circulates, or interference with absorption of the fluid. The signs and symptoms are dependent upon the age of the patient at presentation. In the neonatal period, where the skull is flexible, there is an increase in the head circumference. In older children and adults, there is a gradual development of symptoms of raised intracranial pressure, so that headache, nausea, and vomiting occur, ultimately followed by a deterioration in the level of consciousness. The best treatment is removal of the obstruction if possible, or implantation of a regulatory valve to divert the cerebrospinal fluid to the peritoneal cavity where it is absorbed.

Disruption of the spinal cord, which renders a patient quadriplegic (where all four limbs are affected) or paraplegic (only the lower limbs are affected), depending on the level of injury, is at present incurable—encouraging nerves within the central nervous system to regrow following injury remains a challenging problem for neuroscientists. This is in marked contrast to injuries of peripheral nerves in the limbs which can be sutured or grafted. The nerve beyond the cut will typically die back in the proximal stump but a new nerve can grow (slowly) down the same distal stump after suture. This is a remarkable feat for the nerve cells involved, since in some cases they may be nearly a metre long once fully regrown.

Chapter 5
The head and neck

One of the challenges of studying how the human head evolved and how it develops and functions is its sheer complexity.

Daniel E. Lieberman, Professor of Evolutionary Biology, Harvard University

It is striking that during our embryological development, at about the fourth week, a set of pharyngeal (branchial) arches, separated by pouches, develops. Each pair of arches grows like a collar around the foregut, eventually merging at the midline of the embryo. These arches are thought to be homologous with the gill apparatus of the fish, but in humans they progress to an entirely different fate. It helps the understanding of the anatomy of the neck to consider derivatives from them. Ultimately, the first arch tissues form the upper and lower jaws, two tiny ear bones (the incus and malleus) and the vessels and muscles that supply them. Their nerve supply is via the mandibular division of the trigeminal. The facial nerve (VII) is the nerve of the second arch and all structures that it supplies, which are predominately the muscles for facial expression. Similarly the nerve of the third arch is the glossopharyngeal (IX). The mucous membrane and glands of the back of the tongue are derived from the third arch. The nerves to the fourth and sixth arches are both branches of the vagus (X)—the superior laryngeal and recurrent laryngeal, respectively—which supply the critical functions of the larynx and therefore speech. The fourth and sixth arches become the cartilages of the larynx, while the small fifth arch disappears.

Above these arches, the head is made up of a series of compartments which are the cranial cavity (already described in Chapter 4), the lower jaw (mandible), the orbits, the nasal cavities, and the oral cavity. The lower jaw is shaped like a horseshoe. Each half is L-shaped, consisting of two oblong parts. The horizontal parts of the two sides fuse in the midline during the second year to form the body of the jaw (although in most mammals they remain paired). The alveolar process or upper part of the body of the jaw carries eight teeth in the adult. The rear portion of the jawbone (or *ramus*, which is Latin for *oar*) is an oblong, nearly vertical, flattened plate. It is surmounted by two processes, the head and the coronoid, which are separated from each other by a U-shaped notch: the mandibular notch. The mandibular canal starts at the mandibular foramen and runs through the bone—this narrow structure conveys vessels and nerves from the mandibular division of the trigeminal nerve to the teeth, thus carrying the pain of signals of toothache (Figure 23). The common crowding of human teeth—especially 'wisdom' teeth (third molars), which erupt last—is traceable to evolutionary shortening of our jaw. With easily digestible food available our teeth no longer had to act as heavy-duty food processors and so with the relentless economy of evolution they have slowly shrunk and our jaws have contracted for the same reason.

For movement of the jaw, a joint exists (temporomandibular joint) between the head of mandible and the temporal bone. The strength of this joint depends mainly on the bony conformation and on muscles. A fibrocartilaginous disc caps the head of the mandible and projects forward, dividing the joint cavity into an upper and lower compartment. Movement can occur between the head and disc, and both can move together on the skull, providing a wide range of motion. The heads of the mandible can be palpated by passing a finger into your ear canal and then

Condylar process
Coronoid process
Ramus
Body
Angle
Mental foramen
Mandibular notch
Condylar process
Lingula of mandible
Mandibular foramen
Mylohyoid line
Sublingual fossa
Submandibular fossa
Mental spine

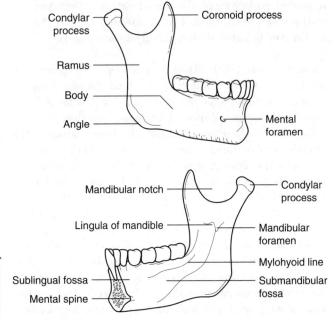

23. The mandible.

opening and closing your mouth, or wiggling the jaw from side to side. The various movements of the jaw are produced by the coordinated activity of the four 'muscles of mastication', medial and lateral pterygoids, temporalis, and the powerful masseter that clenches the teeth.

The eye and orbit

If any one sense can be said to be dominant in humans it must be vision. Primates have been especially visually oriented since our divergence from other mammals. The eye is thought to have developed originally from a simple patch of light-sensitive pigment cells in invertebrates, and such structures are still present in flat worms. The cavities of the orbits are approximately

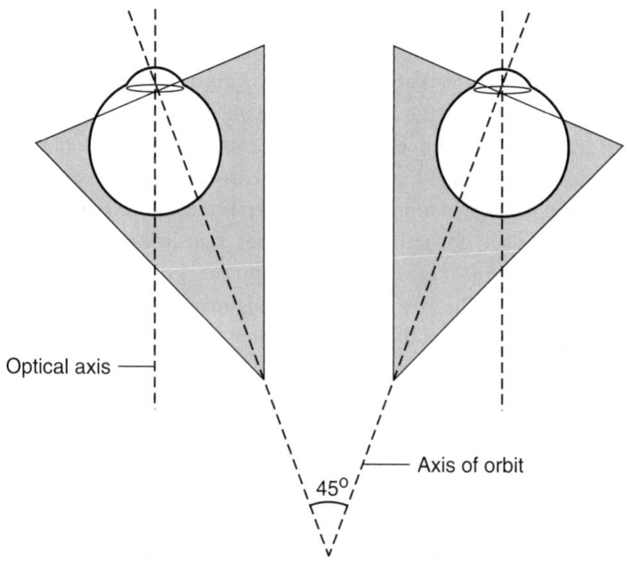

24. The positions of the orbits.

pyramidal in shape. The medial walls are parallel, separated by the nasal cavities, and as a result the axes of the bony orbits lie at 45 degrees to each other. This is of importance for understanding the actions of the six extra-ocular muscles controlling eye movements (Figure 24). The apex is at the optic canal, the site of entry of the optic nerve. In mammals lower than primates there is no bony lateral wall, which is unexpected because in man this wall is strong for transmitting forces from the molar teeth upwards. In many mammals (e.g. the cat or pig) even the lateral part of the orbital margin is missing, although, the lines of force transmission in these animals are distinct from those in humans. The eyeball occupies the front half of the cavity, while muscles and fat largely fill the back half. There are four straight and two oblique muscles which act in a tightly regulated way between the two eyes to provide co-ordinated binocular vision (if this fails, even slightly, this can lead to the phenomenon of 'double vision').

The eyeball contains the light-sensitive retina and like a camera it is provided with a lens system for focusing images and the ability to control the amount of light entering the eye via the iris and pupil. The colour of the eye, blue or brown, is due to the colour of the iris, which is a thin circular muscle controlling the size of the pupil. Most babies of European descent are born with blue eyes. With the passage of time, pigment is deposited, and varying with the amount laid down the colour changes. If little is deposited, the eye remains blue or grey, while at the opposite extreme the eye becomes brown. In Africans the iris is pigmented at birth, and hence in the newborn the iris is not blue.

The wall of the eyeball has three layers. The outer coat is fibrous and consists of the white sclera and translucent cornea. A vascular coat, the black choroid, intervenes between the sclera and innermost nervous layer (the retina). It is black to prevent internal reflections. The retina has several layers and oddly is sensitive to light in the outer layer (this is not the case in most invertebrates and may be considered 'inside out' because the photoreceptors lie behind its ganglion cells). Vision is most acute where rays of light come to focus on the retina at the posterior pole. This part of the retina is the yellow spot or macula. There are two kinds of photoreceptor cells, rods and cones. The centre of the macula, the fovea, is a shallow pit with a particularly high density of cones from which blood vessels are diverted away, thereby decreasing light refraction. The optic nerve pierces the sclera 3 millimetres (mm) to the medial or nasal side of the posterior pole at the optic disc which is the 'blind spot'. The optic disc is clearly visible with an opthalmoscope and is a valuable part of the clinical examination (Figure 25). The appearance of the optic disc on examination varies in different disease processes and can also provide clues as to what is happening in the brain.

The lacrimal gland, situated in the upper lateral part of the orbit secretes tears through a series of ducts into the upper part of the conjunctiva. The conjunctiva is the delicate mucous

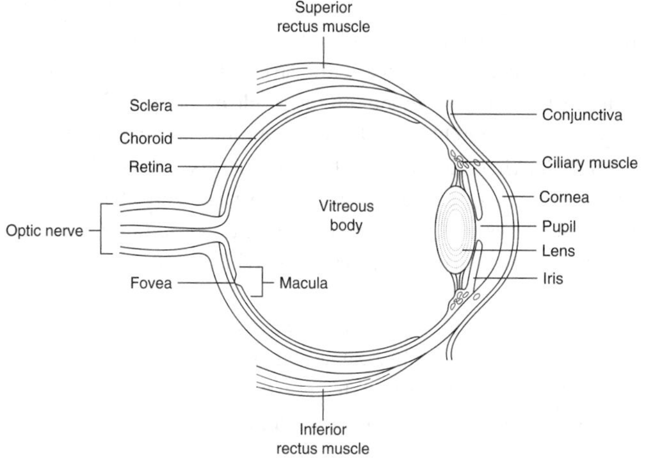

25. Sagittal section of the eye.

membrane lining the inner surface of the eyelids from which it is reflected over the anterior part of the sclera to the cornea. Tears moisten the front of the eye, preventing friction between the eyeball and lids and drying of the corneal epithelium. Tears are drained away through small holes seen near the medial margin of each eyelid. They are collected in the lacrimal sac situated in a small depression on the medial surface of the orbit. This in turn drains via the nasolacrimal duct into the front of the nose, which results in snuffling in the tearful.

When things go wrong

From the vestigial remnants of the second branchial cleft, a cyst (a swelling consisting of a collection of fluid in a sac which is lined by epithelium) may develop. This usually is seen along the line of the sternocleidomastoid muscle, the large muscle clearly seen on either side of the neck. The treatment is by excision.

The retina is at particular risk in relation to injury to the eye. The outer layer of the retina is firmly attached to the choroid but only loosely attached to the inner neural layer, which is largely held in place by the vitreous humour on the inside of the eyeball. High impact forces from blows to the head can sometimes produce enough shearing force to separate the two layers and tear the retina. This is particularly the case in the severely nearsighted because their eyes are longer than average in the anteroposterior direction, which stretches the retina and causes it to be more fragile.

With aging, the lens may become opaque and lose the capacity to transmit light resulting in a cataract. Nowadays removal of cataracts and replacement with a prosthetic lens (silicone or acrylic) is one of the most frequent surgical operations.

The floor of the orbit is its weakest wall, and in blunt trauma such as fist injuries it is often fractured without fractures of the other walls. There may be air in the tissues because the orbital floor is the roof of the maxillary (upper jaw) air sinus (see the next section, which discusses nasal passages). If there is little displacement the fracture heals spontaneously.

The nose

The nasal passages represent a filter through which air must pass en route to the lungs. Beginning in the nasal cavity the air is modified to make it more tolerable to the body. Warming requires a large surface so the walls of the nose are expanded by three conchae, bony scrolls attached to the lateral wall. The superior concha and corresponding region of the septum are covered with olfactory epithelium in which the smell receptors are embedded. Compared with that of other mammals, this is a small area, which corresponds with the reduced emphasis on the sense of smell in man. Both the number and complexity of the conchae (called turbinals in other mammals) are greater in nonprimates.

The nasal cavity connects with the paranasal sinuses. These sinuses are air spaces in several of the facial bones: frontal, ethmoid, sphenoid, and maxilla (the upper jaw). They are bilateral but not symmetrical. Each of the sinuses is lined with ciliated columnar mucous membrane and drains to an opening under cover of the superior or middle concha. The maxillary sinus drains upwards, via an ostium near its roof, a remnant of our time as quadrupeds. The reason for the presence of the paranasal sinuses has been a controversial subject since the time of Galen. Their functions are not clearly known. Several suggestions for the sinuses have been proposed, such as for reduction in weight of the facial bones, to give resonance to the voice, and for thermal insulation for the brain.

Because they represent 'dead ends' in air circulation, sinuses are frequently sites of infection and congestion (Figures 26 and 27).

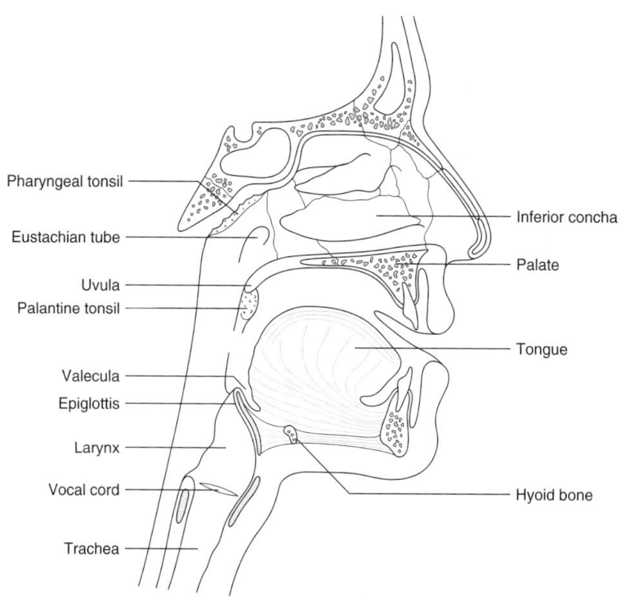

26. Sagittal section of head showing nose, pharynx, and larynx.

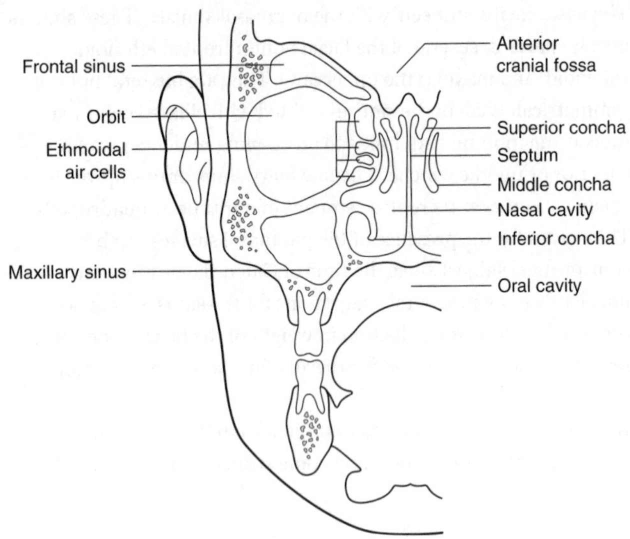

27. Coronal section of face showing orbits, interior of nose, and maxillary sinuses.

The nasal septum separates the right and left nasal cavities. It is made of two bones (ethmoid and vomer) and hyaline cartilage. The septum may not be quite straight and this usually occurs at the junction of bone with cartilage. Just inside the nostrils on the anteroinferior part of the nasal septum is an area where most nose bleeds start, which is called Little's area; it is a site of a confluence of vessels.

Pharynx

The pharynx is a fibromuscular tube that extends from the base of the skull to the lower border of the cricoid cartilage where at the level of the sixth cervical vertebra it is continuous with the oesophagus. The muscles involved are three constrictors: superior, middle, and inferior. The inferior constrictor overlaps the middle and the middle overlaps the superior in telescopic fashion. They are joined together by a posterior midline raphe (or seam). Cricopharyngeus, the lowest part of the inferior constrictor acts as the sphincter (controlling muscle) for the oesophagus.

The air passages of the nasal cavity lead through the openings at the back of the nose, the choanae, into the pharynx. The soft palate acts as a valve to separate the nasal cavity from the oral cavity and to prevent the upward movement of food during swallowing. The pathways for food and air diverge in the lower pharynx (oropharynx), continuing as the larynx and oesophagus, respectively. The epiglottis forms a barrier that deflects food away from the entrance to the larynx during swallowing to prevent choking. In newborns, the larynx is high in the neck and the epiglottis is above the level of the soft palate. Babies can therefore suckle and breathe at the same time. During the second year of life the larynx descends into the low cervical position characteristic of adults.

The pharynx can be described in three parts from top to bottom: nasopharynx, oropharynx, and laryngopharynx. The nasopharynx lies above the soft palate, which cuts it off from the rest of the pharynx during swallowing and thus prevents regurgitation of food through the nose. Two important structures are in this compartment. The nasopharyngeal tonsillar tissue, commonly known as 'the adenoids', is a collection of lymphoid tissue beneath the epithelium of the roof and posterior wall of this region. It forms part of a continuous lymphoid ring with our tonsils (formally described as palatine tonsils) and lymphoid nodules on the back of the tongue. This arrangement produces a ring of lymphoid tissue known as Waldeyer's Ring to act as a guard for the entrance to larynx and oesophagus, protecting against invading pathogens entering from the mouth and nose. The orifice of the pharyngotympanic or auditory tube (Eustachian canal) lies on the side-wall of the nasopharynx level with the floor of the nose. This has the important function of allowing equalization of pressure between the middle and outer ears so as to allow our ear drums to function normally (this is described later in the section on the ear).

The oropharynx is continuous with the oral cavity (mouth). It extends from the uvula of the soft palate (visible in the midline as a structure which hangs down at the back of the throat) above to the tip of the epiglottis below. Its most important contents are the palatine tonsils (our common tonsils). The laryngopharynx extends from the level of the tip of the epiglottis to where the pharynx terminates and the oesophagus begins. The inlet of the larynx lies in front and there is a deep recess on either side of the larynx known as the piriform fossa in which sharp ingested foreign bodies such as fishbones may lodge. The vallecula consist of two depressions on either side of the midline at the base of the tongue, separated by a median fold of mucous membrane. This is a convenient site for the blade of a laryngoscope (an instrument used for passing a tube into the trachea during anaesthesia) as one can pull the tongue forwards.

Larynx

The larynx is a cube shaped box at the upper end of the trachea. It consists of a series of cartilages suspended from the hyoid bone (a U-shaped bone situated at the level of the chin suspended by muscles). The foundation of the laryngeal 'skeleton' is the cricoid cartilage, the only complete cartilaginous ring in the respitatory tract—it articulates with the slender arytenoid and flattened thyroid cartilages. The thyroid has two laminae (thin plates) whose anterior junction constitutes the laryngeal prominence—more prominent in men and known as the Adam's Apple.

Within the larynx, the vocal folds regulate the flow of air out of the lungs and are responsible for voice. A vocal fold consists of a vocal ligament, stretching from the thyroid cartilage anteriorly

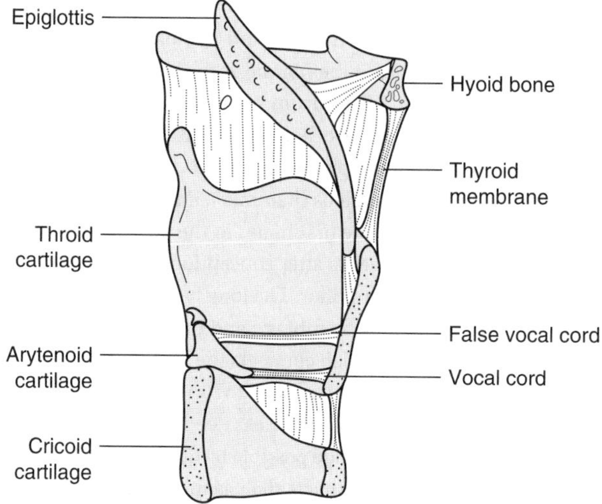

28. Sagittal section of larynx.

to the arytenoid cartilage posteriorly, and a tiny vocalis muscle, which is parallel and adherent to the ligament covered by mucous membranous continuous with the lateral wall. By movements of the cartilages, the free edges of the vocal folds can be shifted subtly medially and laterally to open and close the air passage. If the vocal folds are tensed, air passing between them causes them to vibrate. The vibrations are translated into sound waves as voice. The pitch of the voice corresponds to the degree of tension in the vocal folds (Figure 28).

The ear

The ear has undergone more gross evolutionary transformation in vertebrate history than have other sense organs. The outer ear (pinna) and its external auditory canal collect sounds and transmit them into the cranium. Once sound waves reach the end of the outer ear they are transmitted into the middle ear (an air-filled space within the petrous temporal bone, which is part of the middle cranial fossa) via the eardrum (tympanic membrane). This is a mostly taut fibrous membrane at the end of the bony external auditory canal. Like a drum, the tympanic membrane bows inwards and then deflects backwards with every wave of pressure. The membrane presses against a chain of three middle-ear ossicles: the malleus (hammer), incus (anvil), and stapes (stirrup). They are the first bones in the body to reach adult size at twenty-five weeks after conception. The ossicles function as a bent-lever amplifier. The long handle of the malleus is attached to the tympanic membrane and the stapes is pushed against the oval window, which connects the middle ear to the inner ear. For the tympanic membrane to function optimally the pressure in the middle and outer ear need to be the same. Equalization of pressure is made possible by the pharyngotympanic (Eustachian) tube, which connects the nasopharynx and the middle ear.

The final journey of a sound wave occurs in the cochlear part of the inner ear, which changes the mechanical energy of the amplified sound waves into nerve impulses that are transmitted to the brain. The cochlea (from 'snail') is a spiral-shaped tube of bone with several chambers. Sound waves from the middle ear travel through the fluid of the cochlea every time the stapes knocks against the membrane of the oval window. The lower the frequency, the further the wave moves up the cochlea. As a result it causes vibrations that bend sensory hair cells within the organ of Corti in the central part of the cochlea. This complex organ has a basilar membrane from which project thousands of tiny hair cells (in four rows) that contact an overlying tectorial membrane. When the basilar membrane vibrates, the hair cells bend against the tectorial membrane, stimulating the cell nuclei to send a nerve impulse up the cochlear nerve to the brain. The hair cells are tuned to different frequencies, with higher frequencies received near the oval window and lower ones at increasing

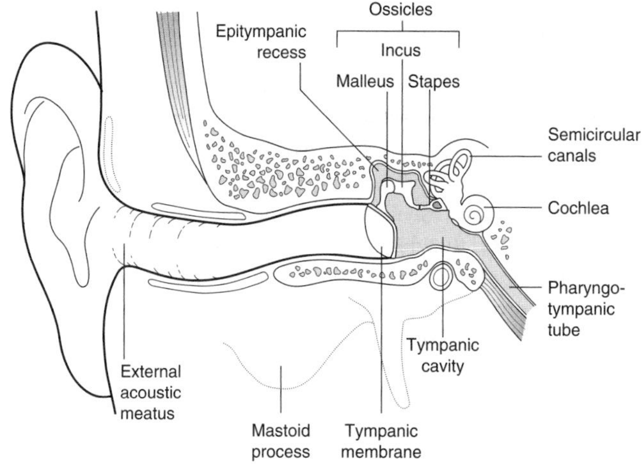

29. The external, middle, and inner ear.

distances from the oval window. There is also a round window, where bone is absent, surrounding the cochlea at the basal end of the tube. One consequence of this structural arrangement is that inward movement of the oval window displaces the fluid of the inner ear, causing the round window to bulge out slightly and deforming the cochlear partition (Figure 29).

The sense of movement is an essential component of all motor activity. Balance involves analysing sensory inputs from the eyes, skin, muscles and joints and the ear also plays a central role here. The vestibule and semicircular canals in the inner ear are the balance organs. The three interconnected semicircular canals are at right angles to each other and can respond to any head movement The vestibule, which consists of the utricle and saccule, responds mainly to the position of the head relative to gravity (static equilibrium), while the semicircular canals react to the speed and direction of head movements (dynamic equilibrium).

When things go wrong

Fractures of the skull may involve the cribriform plate which allows the passage of olfactory nerve rootlets from the olfactory bulb into the nose. This results in leakage of cerebrospinal fluid from the nostrils (rhinorrhea) because of communication with the subarachnoid space. Meningitis is a potential complication. Even blows to the head that do not cause fractures can lead to shearing of the olfactory nerve fibres as they pass through this plate, which results in loss of smell (anosmia). This can diminish the enjoyment of food, influence appetite, and cause weight loss.

Cleft palate with or without cleft lip occurs once in about 2,500 births. There is defective growth of the palatal shelves. Children with cleft palate usually show underdevelopment in the growth of the midface and frequently a reduction of the size of the mandible. There is likely to be difficulty with feeding because food may go into the nose. There is also an increased danger of infections of the middle ear, as the opening of the pharyngotympanic tube is exposed. Cleft lips are usually repaired at 3 months of age and cleft palates at 6 months. These children require careful follow-up during growth, in the form of speech therapy and supervision by ear, nose, and throat as well as orthodontic specialists.

Problems with peripheral hearing can be divided into conductive hearing losses, which involve damage to the outer or middle ear; and sensorineural hearing losses, which stem from damage to the inner ear, most typically the cochlear hair cells or the vestibulocochlear nerve itself. Conductive hearing loss can be due to occlusion of the external ear canal by wax or foreign objects, rupture of the tympanic membrane, or arthritic ossification of the middle-ear bones. In contrast, sensorsineural hearing loss is usually due to congenital or environmental insults that lead to hair cell death or damage to the vestibulocochlear nerve (VIII). In conductive hearing losses, an external hearing aid is used to boost sounds to compensate for the reduced efficiency of the conductive apparatus. The treatment for sensorineural hearing loss is more complicated and invasive. Conventional hearing aids are useless, as no amount of mechanical amplification can compensate for the inability to generate or convey a nerve impulse from the cochlea. However, if the auditory nerve is intact, an electronic device, a cochlear implant, can be used to partially restore hearing. The electrode is inserted into the cochlea through the round window and positioned along the length of the basilar membrane.

Chapter 6
The trunk—abdomen and pelvis

The thoracic and abdominopelvic cavities are closed to the outside. Viscera located within these closed cavities often have cavities open to the outside e.g. digestive tract, uterus, bladder.

Larence M. Elson, Anatomist, University of California School of Medicine, San Francisco

The trunk is the body without the limbs. The skeletal component consists of the ribs and sternum, the spine, and the pelvis. The organs of the cardiovascular and respiratory systems have been covered in Chapter 3. The gastrointestinal and genitourinary tracts are within the abdominopelvic cavity.

The dorsal musculature acts to extend the spine while the ventral muscles are flexors of the spine. The spinal extensor group is known as the 'erector spinae', which extends from the pelvis to the skull. There are three columns of muscle fibres on each side of the vertebral column. Since individual bundles within these columns arise and insert from each segment (vertebra), the erector spinae should be considered a muscle group rather than distinct muscles. The muscles on each side laterally flex the spine in that direction, but acting together the erector spinae muscles extend the spine. Deeper and smaller groups of muscles help to stabilize and rotate the intervertebral joints.

The abdominal wall is not protected by bone and is maintained entirely by its muscular layers. The abdominal wall consists of three layers of muscle. The fibres in each layer run in a different direction relative to those of the other two layers. The resulting muscular wall resembles a sheet of plywood in which greater strength is obtained by alternating mutually perpendicular grains. Criss-crossing fibres also narrow the waist of the abdominal wall where there is no skeletal support. The orientation of these muscles is similar to those of the thoracic wall, which suggests some continuity in development. Each of these muscles ends partly in a broad sheet-like tendon or aponeurosis, which is anchored in the midline. The muscles are from outside-in: the external oblique, internal oblique, and transversus abdominis. The rectus abdominis is a midline muscle strengthened by the insertions of the anterolateral muscles. It is a powerful flexor of the spine. There are three or four transverse tendinous intersections in the rectus abdominis, which are indicative of its segmental origin. These become very obvious when the muscle hypertrophies as the apparent 'washboard' effect or 'six pack' clearly seen in thin, fit athletes. All these muscles create and regulate pressure in the abdomen. The pressure on the lumbar vertebrae supports the spine as, for example, when lifting a heavy object. Abdominal pressure restores the diaphragm to its resting position after expiration (breathing out) and provides the power for coughing and blowing, as well as during defecation and childbirth. The muscles described are commonly called 'core muscles' by fitness trainers because they stabilize the trunk in all sports.

There are, however, two main sites of weakness in the anterior abdominal wall. These are the inguinal canals, which, during early development, provided passage for the testes to reach the scrotum in the male and for the round ligament to reach the labia majora in the female. A hernia is the protrusion of an internal organ (or viscus), or part of one, through an abnormal opening in the walls of its containing cavity—such as through these inguinal weaknesses. Inguinal herniae are common in men and less so in women.

The 'abdominopelvic cavity' is the space enclosed by the bones, muscles, and fasciae of the abdominal and pelvic walls from the diaphragm above to the pelvic floor ('pelvic diaphragm') below. This cavity is divided at the pelvic inlet into the abdominal cavity and the pelvic cavity. These two divisions are set nearly at right angles to each other (Figure 30). The pelvis is frequently described in two parts. The false pelvis refers to the space between iliac blades, while the true (anatomical) pelvis lies below the pelvic brim. The bony pelvis of males and females differ because of the need for childbirth. All diametres of foramina are absolutely greater in women than in men.

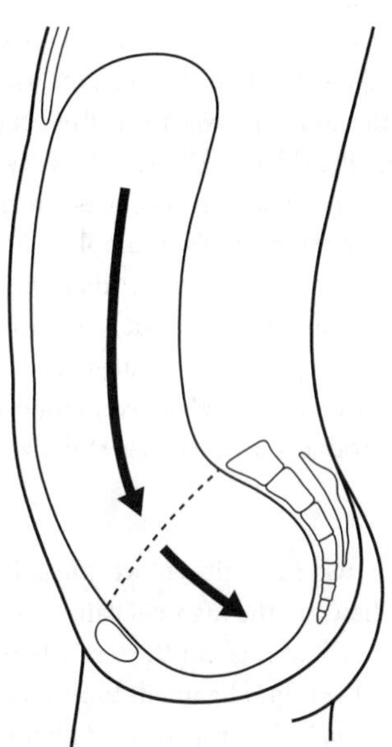

30. Relationship of abdomen and pelvis.

The gut and abdominal cavity

The early gastrointestinal tract is oriented longitudinally—head to toe—and is delicately suspended from the surrounding abdominal walls by two large strings of blood vessels—or 'mesenteries'—in front and behind. The primitive gut tube consists of the foregut, midgut, and hind gut. Each section has its own blood supply—and this arrangement remains throughout development into the adult. These arteries supply blood from the aorta behind: the coeliac artery supplies the foregut, the superior mesenteric artery the midgut, and the inferior mesenteric artery the hindgut.

The foregut gives rise to the lower end of the oesophagus, the stomach, and the first part of the duodenum. The midgut develops into the latter part of the duodenum, the small bowel, ascending colon, and first two-thirds of the transverse colon. This is a large amount of bowel to be packed into the abdomen as it grows, and evolution has provided an elegant mechanism to allow this. Rapid growth of the gastrointestinal system results in a loop of midgut pushing forward out of the abdominal cavity into the umbilical cord in an extended U-shape. While extended in this way, the two limbs of the midgut rotate counterclockwise around the superior mesenteric artery. The bowel is then returned to the abdominal cavity and the part of the loop that is destined to become the caecum descends into its definitive position in the right iliac fossa.

The final parts of the colon, the rectum, and upper part of the anal canal develop from the hindgut.

The abdominal cavity is lined throughout by a thin sheet, or peritoneum, which clings to the organs (or viscera) as 'visceral peritoneum'; and lines the surfaces (anterior and posterior) as parietal peritoneum. At the front, a large, fatty peritoneal fold hanging from the stomach and transverse colon forms the protective greater omentum, also known as 'the abdominal

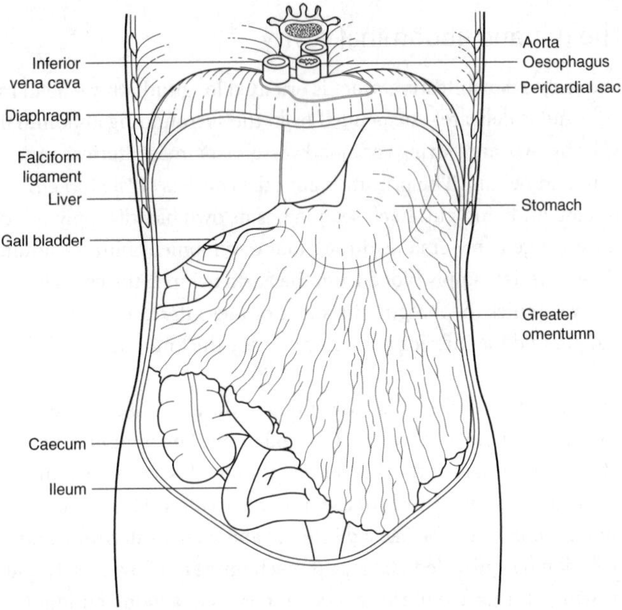

31. The greater omentum.

policeman' (Figure 31). This moves passively due to contractions of the bowel or it may be pushed by the abdominal wall into an area of bowel damage which it serves to protect, rather like a bandage. An acutely inflamed appendix is often found wrapped in omentum, which prevents the development of general peritonitis. At the back, many parts of the intestine in the adult remain connected to the posterior abdominal wall by mesenteries, which convey vessels and nerves and are also coated in peritoneum. Structures which are suspended on mesenteries, such as the small bowel, are freely mobile.

The gastrointestinal tract comprises the oesophagus, stomach, and small and large intestines (Figure 32). Their primary function is to provide the body with a continual supply of water, electrolytes, and

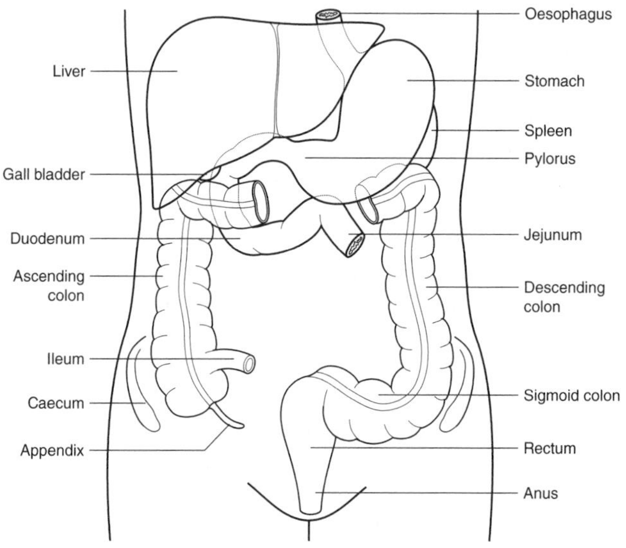

32. Stomach and small, and large bowel.

nutrients, but before this can be achieved food must be moved along at an appropriate rate for the digestive and absorptive functions to take place. Two basic types of movement occur, mixing movements and propulsive movements or 'peristalsis'. Food which has been swallowed first passes from the pharynx into the oesophagus. The oesophagus is a tube that is about 25 centimetres (cm) long with the last 3–4 cm located in the abdominal cavity below the diaphragm. Its sole function is to transmit food to the stomach quickly. It has no digestive action. The stomach is a muscular chamber that is specialized for mechanical churning of the food and for some chemical digestion. The mucosa (inner lining) has long folds, the rugae, which enable the stomach to expand when full. The stomach contents are highly acidic to activate digestive enzymes and to kill ingested bacteria. The stomach lies to the left of the midline, with its exit, the pylorus, being located to the right of the midline. It passes food on to the small intestine, which is subdivided into three parts: duodenum, jejunum, and ileum.

The duodenum is a specialized horseshoe-shaped portion of the small intestine that has lost its primitive mesentery. It begins at the pylorus and ends at the doudenojejunal junction. Having no mesentery, the duodenum adheres to the structures on the posterior abdominal wall of the abdomen. The jejunum and ileum are suspended on mesenteries and freely mobile. They extend from the duodenojejunal junction to the right iliac fossa (see Figure 32) where the terminal ileum opens into the large intestine at the caecum. The small intestine is the most critical region of the gut and is generally between 5.5–6 metres long. It is where most of the enzyme secretion and chemical digestion and nearly all the nutrient absorption occurs. While the jejunum and ileum have distinctive characteristics, the changes occur gradually. Typically the jejunum is of greater calibre, has a thicker wall, and its arteries are more closely packed. The ileum on the other hand, is thin walled, contains prominent aggregations of lymphoid tissue called Peyer's patches, and its arteries are more widely spaced. The Peyer's patches are important for recognition and rejection of harmful gut bacteria (or pathogens).

The caecum is the blind cul-de-sac situated below the ileocaecal orifice. The worm shaped (vermiform) appendix opens into the inferior end of the caecum. It is about the length of an adult's little finger. It has no digestive function and is notorious for becoming infected and inflamed. However it is now recognized as being not simply vestigial, as its walls contain lobules of lymphatic tissue which, like the Peyer's patches, help in immune surveillance. The appendix has therefore been called the 'abdominal tonsil'.

The colon stretches round the outer edges of the abdominal cavity and the main sections describe this roundabout passage. The ascending colon ascends from the right iliac fossa to the undersurface of the liver where it makes a bend, the hepatic flexure, and becomes the transverse colon. This extends across the abdomen to the under surface of the spleen, where it bends again at the splenic flexure. From here the descending colon extends to the pelvic brim where it becomes the sigmoid (pelvic) colon. The sigmoid colon passes (in a looping S-shape, hence its name) to the middle of the sacrum where it becomes the rectum. The lower portion of the rectum and anal canal pass through the floor of the pelvis and, in the perineum, opens to the exterior through the anus.

Unlike the small intestine the outer longitudinal muscle coat of the large intestine does not form a complete coat, instead it is arranged in three narrow bands, the taenia coli, which being shorter than the gut causes it to be gathered up into sacculations or small segments. Peritoneal tags of fat (appendices epiploicae), hang from the large gut throughout its length. Size alone does not necessarily distinguish large gut from small gut, as the descending colon has a calibre less than that of the small gut, but functionally they are quite distinct. Most of the nutrients should have been absorbed by the ileum before bowel contents enter the colon. Any further absorption, mostly of water, is completed in the proximal or absorbing colon while the distal colon functions principally for storage and is called the storage colon. The proximal colon also houses a considerable population of symbiotic bacteria. This live-in colony of microbes consists of hundreds of individual species. It is only recently that new genomic techniques have opened the door to our gut microbiome. Understanding how it varies between individuals is important. Scientists believe that the pattern of the inhabitants of our gut might help explain why some people develop metabolic disorders, and why some put on weight when others remain thin.

Liver

The liver is the second largest organ in humans. It is exceeded only in size by the skin. It has a major role in metabolism and all vertebrates have a liver. It is located in the upper right quadrant of the abdominal cavity protected by the rib cage. The liver is one of the most well-vascularized organs of the body (it receives 25 per cent of the cardiac output), but it has an unusual blood supply. First, it is supplied by arterial blood via the hepatic artery, which is a large branch of the coeliac artery. However, unlike most organs it is also supplied with venous blood via the portal vein that runs directly from the intestines and spleen to the liver delivering absorbed products of digestion—except for fat, which passes via the lymphatics into the blood. The liver is divided anatomically into a larger right and smaller left lobe, but these have identical functions and indeed half the liver can be removed in some situations (hemihepatectomy) with liver capacity maintained.

Bile is produced continuously by the cells of the liver and secreted into the hepatic duct system. The bile accumulates in the gallbladder whose main function is to serve as a reservoir for bile. The gall bladder is a pear-shaped stretchable sac which lies on the underside of the right lobe of the liver. A gall bladder is

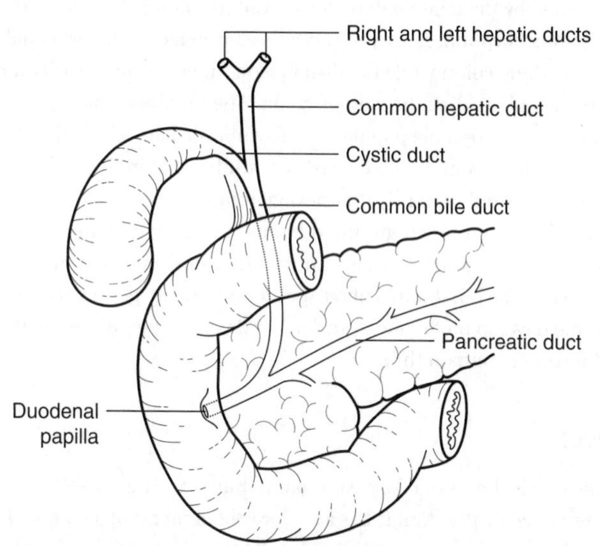

33. The biliary tract. The gallbladder is connected by the cystic duct to the common hepatic duct to form the common bile duct.

present in most species of fish and in orders of vertebrates higher than fish. The duct draining the gall bladder, the cystic duct, unites at an acute angle with the common hepatic duct, which is formed from the right and left hepatic ducts of the right and left lobes of the liver to form the common bile duct. This passes behind the duodenum and the head of the pancreas to join the second part of the duodenum, which it enters through a small raised projection together with the duct draining the pancreas. The walls of the gall bladder are muscular and bile is only released when they contract, in response to signals that a flow of bile is needed for the digestion of food. Bile salts help the break down (emulsification) of fats (Figure 33).

The pancreas (sweetbread)

Embryologically and in many adult vertebrates, the pancreas is a paired organ, with two parts that lie in front of and behind the duodenum. In humans, however, the ventral (anterior) pancreas rotates during development and fuses with the pancreas behind the duodenum to form a single organ. This origin explains the irregular shape of the gland and the variable appearance of one or two ducts. The pancreas is fish-shaped with a head, body, and tail, all lying behind the peritoneal sac. The head lies within the C-shaped curve of the duodenum, the body crosses the midline, and the tail reaches the spleen. The main duct resembles a herring bone in that small ducts spring from the main duct, which is straight. This duct empties the digestive secretions of the pancreas into the duodenum together with the common bile duct. The opening of these two ducts (the ampulla of Vater), is guarded by a sphincter to prevent the gut contents moving backwards into these critical organs.

The pancreas has two functions. It acts as an 'exocrine' gland, and each day discharges an enzyme-containing juice into the gut for digesting all three major types of food: proteins, carbohydrates, and fat. It is also an 'endocrine' gland, producing and secreting hormones directly into the blood stream to act at distant target sites in the body. The endocrine secretion is formed by tiny clusters of cells, the islets of Langerhans, which produce insulin and other important hormones that are vital for carbohydrate metabolism. Depletion of insulin results in diabetes.

The spleen is the largest of the lymphocyte producing organs. It lies next to the pancreas in the left-upper quadrant of the abdomen, deep to the ninth, tenth, and eleventh ribs (its long axis follows the tenth rib on the left). The spleen is not essential for life but its removal results in a markedly increased susceptibility to infections and septicaemia. It has a capsule which allows it to expand and contract; however the capsule is thin and as a result the spleen is the commonest intra-abdominal viscus to be ruptured by blunt trauma, such as falling off a horse.

When things go wrong

Meckel's diverticulum occurs in 2 per cent of the population and represents the remnant of the vitello-intestinal duct which connects the primitive midgut and yolk sac. It lies on the opposite border of the bowel to the mesentery, and commonly occurs 60 cm from the ileocaecal junction. The mucosa lining the diverticulum may contain islands of gastric mucosa with acid-secreting cells. Peptic ulceration of adjacent intestinal epithelium may then occur with haemorrhage or perforation. This may mimic acute appendicitis.

Exomphalos is the persistence of the midgut herniation at the umbilicus after birth. Umbilical hernia of infants occurs through a weak umbilical scar but disappears spontaneously in 90 per cent of cases. Rarely the two developing segments of the pancreas fuse and completely surround the second part of the duodenum as an annular pancreas, which may produce duodenal obstruction.

The adrenal glands, kidneys, and gonads

These paired structures all develop in the tissue on the posterior abdominal wall, behind the peritoneum, although they move differently during development. The adrenal glands remain in situ, while the reproductive organs (testes and ovaries) descend and the kidneys ascend. The adrenals, paired endocrine glands, are crescentic structures which overlap the upper ends of the kidneys. Each gland is composed of two parts: an outer golden cortex and an inner vascular medulla. The different parts of the glands are linked anatomically but are quite distinct functionally; the adrenal medulla is functionally related to the sympathetic nervous system rather than other endocrine tissue. The glands lie at the level of the twelfth thoracic vertebra. Preganglionic sympathetic nerve fibres pass, without synapsing, all the way from the intermediolateral horn cells of the spinal cord, through the sympathetic chains, and the splanchnic nerves, and finally into the adrenal medulla. There they end on special cells that secrete epinephrine (adrenaline) and norepinephrene (noradrenaline), two complementary hormones, directly into the blood stream. This constitutes an amazingly effective response for emergencies (the autonomic nervous system is discussed in Chapter 4). The adrenal cortex secretes an entirely different group of hormones called corticosteroids. These hormones are all synthesized from the steroid cholesterol and they all have similar chemical formulae. Minimal differences in their molecular structures confer very different and important functions (Figure 34).

The kidneys perform two major functions: they excrete most of the end-products of bodily metabolism; and, second, they control the concentration of most of the constituents of the body fluids. The two kidneys together contain about two million nephrons, each of which is capable of forming urine by itself. The lower poles of the kidney lie at the level of the lumbar third or fourth intervertebral disc and the upper poles at the level of the twelfth

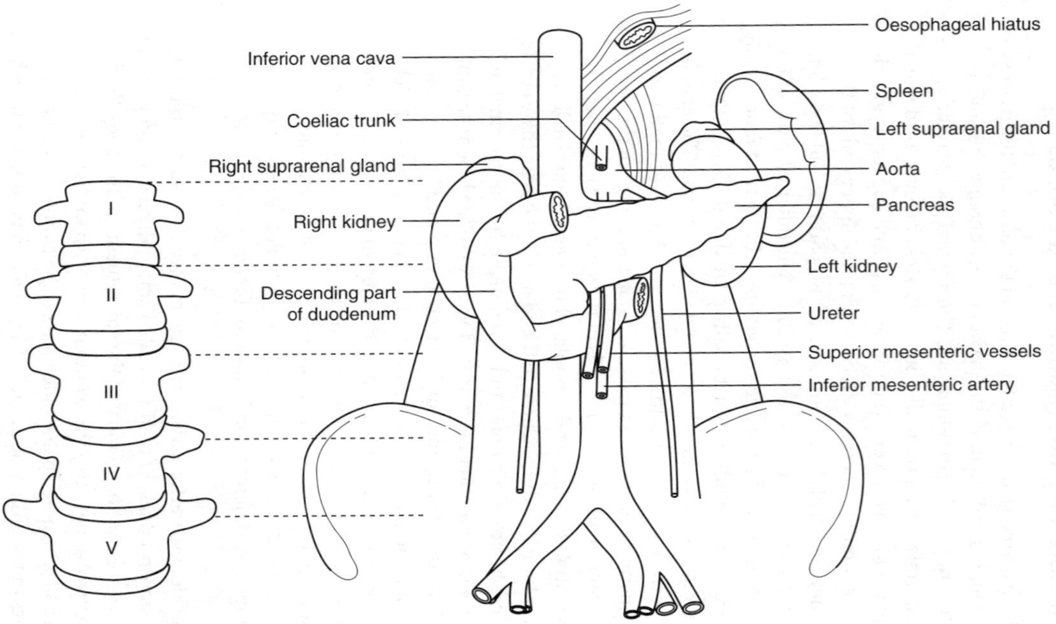

Oesophageal hiatus

Inferior vena cava

Spleen

Coeliac trunk

Left suprarenal gland

Right suprarenal gland

Aorta

I

Pancreas

Right kidney

II

Left kidney

Descending part
of duodenum

Ureter

III

Superior mesenteric vessels

Inferior mesenteric artery

IV

V

34. Kidneys, suprarenals, and pancreas.

thoracic vertebra. The kidneys lie in the retroperitoneal fatty tissue. In the midst of this fatty tissue there is a tough membrane, the renal fascia, which splits to enclose the kidney and a certain quantity of fat—the perinephric fat or fatty capsule. The two layers of fascia do not blend below, allowing for the kidneys to move downward with the diaphragm during inspiration.

The blood supply of the kidney (about 20 per cent of the cardiac output) is peculiar in that all, or almost all, of the blood passes through the capillaries of the nephrons, where it is purified, creating the urine, before it passes through a second set of capillaries from which it nourishes the kidney substance. The renal arteries, one for each kidney, arise from the sides of the aorta just below the superior mesenteric artery. The left renal vein is longer than the right. They both drain directly into the inferior vena cava (main venous channel of the abdomen).

The urine formed from the renal capillaries drains into the renal pelvis. The latter is continuous with the ureter that connects the kidney to the bladder, which lies, in the adult, behind the pubic symphysis (it is higher in the child but descends with growth). Urine collects in the bladder, which can be voluntarily voided from time to time via the urethra. In the female, the urethra is short (4 cm) opening just in front of the vaginal orifice. In the male, the urethra is 20 cm and passes through the prostate gland to the external urethral orifice of the glans penis. Micturiton is the process by which the urinary bladder empties when it becomes filled. The bladder progressively fills until the tension in its walls rises above a threshold value at which time a nervous reflex occurs that either causes micturition or, if voiding is inconvenient, at least causes a conscious desire to urinate.

The pelvis and perineum

The pelvis ('basin' in Latin) is the part of the trunk that is below the abdomen, and is the area of transition between the trunk and

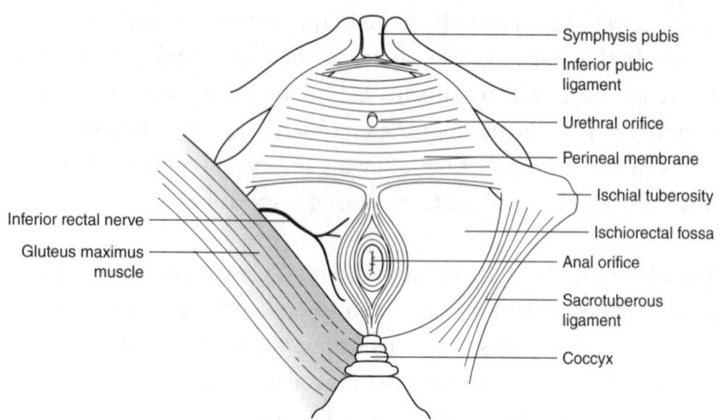

35a. Urogenital and anal triangles of perineum.

lower limbs. The perineum is the region of the trunk below the pelvic diaphragm containing the outflow from the rectum and bladder. It is a diamond shaped space (Figure 35a). The anterior half of this diamond is the urogenital triangle and the posterior half the anal triangle. In the embryo the alimentary tract ends in a blind receptacle, the cloaca. In reptiles and birds the cloaca opens on to the skin surface through an orifice guarded by a sphincter of striated (voluntary) muscle. In mammals including man, during development, a septum divides the cloaca into an anterior or urogenital part and a posterior or intestinal part. This helps to explain why a single nerve, the pudendal nerve, supplies all the local muscles and sensation and its companion artery, the pudendal artery, nourishes the entire territory, and also why the bladder and rectum have a common autonomic nerve supply.

The pelvic diaphragm consists of the levator ani muscles, which are a pair of saucer-like slings of muscle attached to the inner surfaces of the pelvic cavity (Figure 35b).

This muscular layer separates the pelvic cavity from the perineum, for which it forms the roof. In the dog the posterior part of levator ani (coccygeus) wags the tail, but in man it is largely transformed

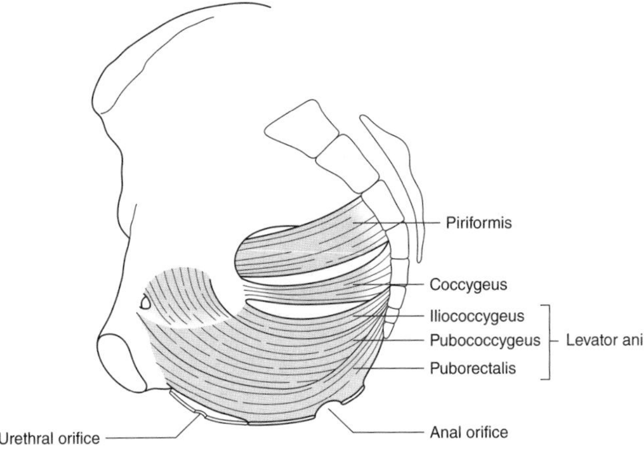

35b. Component of pelvic diaphragm.

into the sacrospinous ligament with some fleshy fibres on the pelvic surface of the ligament. The pelvic diaphragm is perforated by the urethra and anal canal and in the female also by the vagina. Although composed of multiple muscle components, the levator ani muscles act as a unit to aid in control of faecal continence, micturition, and to prevent prolapse of pelvic viscera. The perineal membrane is a sheet of strong fascia, which stretches between the two sides of the pubic arch. Above it, muscle fibres (transversus perinei profundus) run transversely and other fibres (sphincter urethrae) in part encircle the urethra and in part decussate (split) and embrace it. The space deep to the perineal membrane is called the deep perineal pouch. Its contents include the membranous urethra, the urethral sphincter, and the bulbourethral glands, which lubricate the urethra for intercourse.

The rectum and anal canal are the terminal parts of the large intestine. The rectum begins where the sigmoid colon ceases to have a mesentery, which is in front of the third piece of the sacrum (Figure 36). It is 12 cm in length. The rectum is straight in lower mammals (hence its name) but is curved in man to fit the sacral

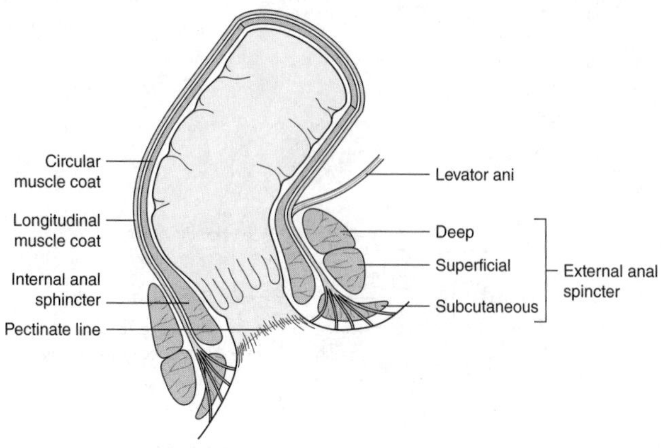

36. Rectum and sphincters.

hollow. It continues the curvature of the sacrum and coccyx downwards and forwards for 4 cm beyond the coccyx, and there at the apex of the prostate makes a right-angled bend and becomes the anal canal. The anorectal flexure of the anal canal that occurs as the gut perforates the pelvic diaphragm (levator ani) is an important mechanism for faecal continence. The rectum has three lateral curvatures with associated projections of shelves of mucous membrane and circular muscle into the lumen, left, right, and left from above downwards (valves of Houston), which need to be negotiated when imaging instruments are introduced from below.

The anal canal extends from the anorectal junction, which lies above the level of the puborectal sling and the sphincters, to the anus below. Its function is to remain closed, except when the colon and rectum are expelling their contents. Accordingly it is surrounded throughout by two sphincters—an external voluntary and internal involuntary one, which is merely the thickened lower end of the circular muscle of the gut. Developmentally the anal canal has a twofold origin, an ingrowth from the skin (proctodeum) and a downgrowth of the hindgut towards the surface. During development, the anal membrane is the partition that divides the proctodeum from the hindgut. It disappears leaving only a wavy whitish line—the pectinate or dentate line—to mark the mucocutaneous junction (i.e. between gut and skin). The upper part of the anal canal has five to ten permanent longitudinal folds of mucous membrane, known as the anal columns, whose lower ends are united by semilunar folds, the anal valves, at the pectinate line. These columns contain the terminal branches of the superior rectal artery and vein. The superior, middle, and inferior rectal arteries supply the rectum and anal canal. These vessels have corresponding veins. The common disorder piles (or haemorrhoids) is a prolapse of rectal mucosa containing dilated veins and occurs because of increase of pressure in this valveless system of veins due to chronic constipation or pregnancy. Because of the presence of abundant arteriovenous anastomoses, bleeding from internal piles is characteristically bright red.

The genital systems

One of the striking features in the development of the reproductive system is the initial lack of sexual differentiation. One might expect that male and female reproductive mechanisms, which are so dissimilar in their adult forms, would be sharply differentiated from one another from their earliest appearance. However, embryos exhibit gonads which at first give no evidence as to whether they are destined to develop into ovaries or testes. There is thus a common starting point for the later developmental changes into either gender.

The penis (Latin for 'tail') is composed of three fibroelastic cylinders, the right and left corpora cavernosa and the corpus spongiosum, which are filled with erectile tissue and are enveloped in fasciae and skin. The corpora cavernosa fuse with each other in the median plane, except behind where they separate to find attachments to the adjacent pubic arch. They are the support of the corpus spongiosum, which lies below and between them. The corpus spongiosum is traversed by the urethra. It is expanded in front as the glans penis, which fits onto the united corpora cavernosa, and expanded behind, while the bulb of the penis is fixed to the perineal membrane. The skin and fascia of the abdominal wall and scrotum are prolonged over the penis as a series of very loosely laminated envelopes which end as the foreskin or prepuce. The erectile tissue of the penis is supported by three paired arteries. When the penis is flaccid, the arteries to the corpora cavernosa are coiled, restricting blood flow, and they are known as the helicine arteries of the penis. These arteries straighten when an erection of the penis occurs. There are also three muscles for the penis. One of these, bulbospongiosus, functions as a sphincter that empties the bulb and base of the spongy urethra.

The testis is an ovoid gland which is around $5 \times 3 \times 2.5\ cm^3$ in size. It has a tough fibrous, outer coat, the tunica albuginea, which is comparable to the sclerotic white outer coat of the eyeball. Septa (walls) of connective tissue project into the testis making a number of compartments. A number of highly coiled seminiferous tubules are packed into each compartment and these produce the sperm. Sperm are stored and mature within the epididymis, which is attached to the upper pole and posterior border of the testis. It is divided into a head or upper part (where sperm enter), a body or intermediate part, and a tail or lower part (where sperm are stored). The testis is supplied by the testicular artery, a direct branch of the abdominal aorta, which arises just below the renal arteries. It joins with the artery to the vas deferens and epididymis, which thus provides an additional blood supply for safety. The testes, in addition to producing sperm, secrete several male sex hormones, which are collectively called androgens. However, testosterone is so much more abundant and potent than the others that it is the significant hormone of the male.

During development the testes descend from the posterior abdominal wall. A mesenchymal strand, the gubernaculum testis, extends from the lower end of the developing testis along the course of its descent to blend into the scrotal fascia. The exact role of this structure in the descent of the testis is not known. It has been suggested it acts as a guide (gubernaculum means 'rudder') or that its swelling dilates the inguinal canal and scrotum. The testes traverse the inguinal canal during the seventh prenatal month and reach the bottom of the scrotum after birth. They are followed by an outpouching of the peritoneum—the processus vaginalis. At birth this is obliterated leaving the testis covered by a fine layer known as the tunica vaginalis. When the processus remains patent, because the two layers of peritoneum do not fuse, it allows the bowel to enter the scrotum from the abdomen resulting in a congenital inguinal hernia. Fluid may also travel down this to form a hydrocele, a fluid-filled sac surrounding the testis. This can occur spontaneously (primary) or in later life, secondary to testicular disease.

The scrotum is the bag of skin and subcutaneous tissue in which the two testes lie. The skin of the scrotum forms a single pouch, but the subcutaneous tissues form a right and left pouch with a common midline partition. The subcutaneous tissue contains a sheet of involuntary muscle, the dartos muscle. The fibres of the dartos adhere to the skin and cause it to wrinkle when cold and thus assisting the cremaster muscle which lies in the spermatic cord to draw the testis upwards in the scrotum. In a warm environment, such as a hot bath, the cremaster muscle relaxes and the testis descends deep into the scrotum. Both responses occur in an attempt to regulate the temperature of the testis for spermatogenesis (formation of sperm), which requires a constant temperature approximately 2 degrees below core temperature. The testis sometimes fails to leave the abdomen, an undescended testis, or may be found at another site, an ectopic testis, leaving the scrotum empty. In contrast, in the elephant the testis is retained in the abdomen; in certain rodents it descends periodically and then returns to the abdomen; in the pig it descends to the perineum; and in marsupials it becomes prepenile.

The prostate, a fibromuscular and glandular organ, surrounds the urethra between the bladder and the perineal membrane. Prostatic fluid, a thin milky fluid, provides approximately 20 per cent of the volume of semen (a mixture of secretions produced by the testes, seminal glands, prostate, and bulbourethral glands that provides the vehicle by which sperm are transported) and plays a role in activating the sperm.

The vas deferens is the duct that conveys spermatozoa from the testis to the urethra. It has a thick muscular coat and a capillary lumen and it is usually described as feeling firm like a whip cord. It is this tube that is tied off in a vasectomy. The vas passes from the scrotum via the inguinal canal to the back of the bladder. Obliquely placed seminal glands converge at the base of the bladder, where each of their ducts merges with the vas to form an ejaculatory duct. The two ejaculatory ducts immediately enter the posterior aspect of the prostate, running through the gland to open in the prostatic urethra. The urinary and reproductive tracts converge at this point (Figure 37).

In the female the ovary is shaped like a testis but is only half the size. The pits and scars on its surface mark the sites of previously

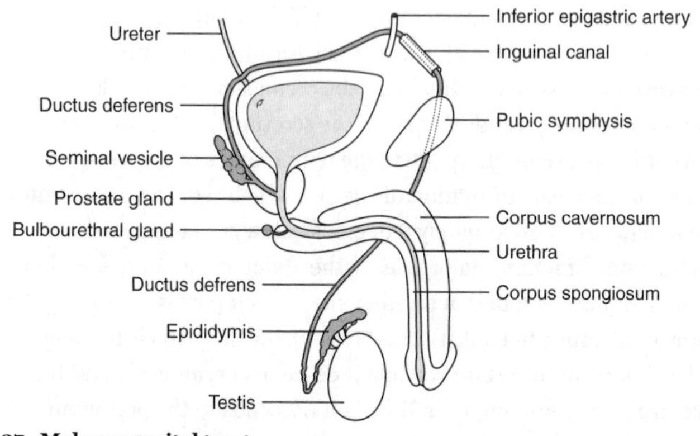

37. Male urogenital tract.

shed ova. The ovary is suspended by a ligament from the pelvic side wall which also conducts the ovarian artery, vein, lymph vessels, and nerves to the ovary. The ovarian vessels duplicate the course of the testicular vessels until they plunge into the pelvis. The round ligament takes the same course as the vas deferens in the male but ends in the labium majus, which is the homologue of the scrotum.

The uterus is a pear-shaped organ 7.5 cm in length, made up of the fundus, body, and cervix. The uterine (Fallopian) tubes enter the upper angles (the cornu) above which lies the fundus. The body of the uterus narrows to a waist termed the 'isthmus', continuing into the cervix, which is embraced about its middle by the vagina. The narrow cavity of the uterus communicates with the cervical canal, which in turn opens into the vagina. Around the cervix there is a circular gutter, created by the folds of the vagina; the rear part of this gutter is adjacent to the recto-uterine pouch, the lowest part of the peritoneal cavity, and thus well-placed as a site for drainage of pus which might occur in the pelvis during disease. The uterus is normally tipped forward so that its weight is born largely by the bladder. Further passive support for the uterus is provided by ligaments with active support from the levator ani. The uterus changes in size with the different stages of a lifetime. During pregnancy the uterus increases in size due to the enlargement of individual smooth muscle cells, which it has been estimated can by simple hypertrophy (enlargement), increase their bulk eightfold.

The uterine or Fallopian tubes are about 10 cm long. Each uterine tube has an expanded trumpet-shaped end (the infundibulum), which curves around the end of the related ovary. The rim of this is covered with small finger-like projections (fimbriae), one of which is long enough to reach the ovary. The lumen of the tube opens into the peritoneal cavity (thus there is a potential link between the peritoneum and the outside world through these tubes, the uterus, and the vagina). After ovulation, the unfertilized

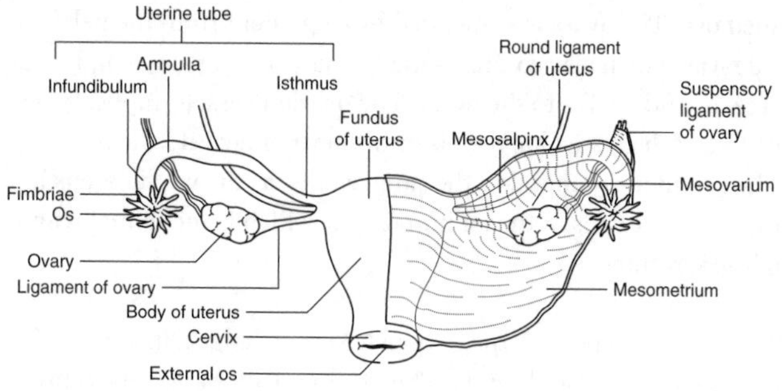

38. Female pelvic organs.

egg is gathered by the fimbria of the uterine tube and the egg passes into the tube where it may be fertilized. The fertilized egg then normally continues into the uterine cavity and implants in the uterine wall (Figure 38).

Interestingly, the perineum, in the foetal condition, is similar in both sexes. However, following development these structures diverge, although there are equivalent structures in both sexes. The female penis or clitoris is diminutive and not traversed by the urethra. It comprises two corpora cavernosa clitoridis and a glans clitoridis just as in the male. The developmental equivalent of the scrotum is split into right and left labia majora. Each labium majus is a broad ridge lying lateral to the vaginal opening and covering a long fingerlike process of fat. This process extends backward from a median skin-covered mound of fat, the mons pubis, which lies over the pubic bone.

The bulbospongiosus muscle (also present in the male) splits into right and left parts resulting in a sphincter of the vagina. The hymen is a thin membranous fold that surrounds the vaginal orifice which is perforated to allow discharge of menstrual fluid. At first coitus the hymen tears and after childbirth no hymenal tissue remains apart from hymenal tags in some cases (Figure 39).

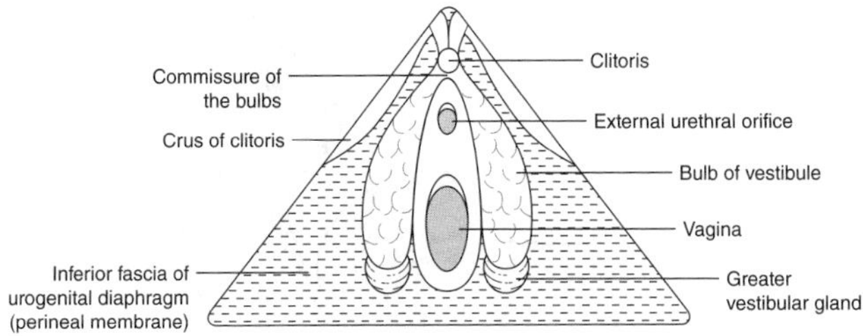

39. External genitalia of the female.

The peritoneum lining the abdominal cavity continues down into the pelvic cavity, but folds back onto the pelvic organs and thus it is separated from the pelvic floor. The region above the bladder is the only site where the peritoneum is not firmly bound to the underlying structures. This allows the bladder as it fills to extend up into the loose areolar tissue between the peritoneum and the anterior abdominal wall. In the female, as the peritoneum reaches the rear of the bladder it passes across to the uterus. It then descends behind the uterus on to the posterior vaginal wall before climbing back onto the rectum. The 'pocket' thus formed between the uterus and rectum is the recto-uterine pouch (of Douglas) often described as the most inferior extent of the peritoneal cavity. This is a site where pus may collect if there is disease in the pelvis (Figure 40).

When things go wrong

Unilateral absence of a kidney is relatively common and a single kidney can typically provide sufficient function. Pelvic kidneys result from failure of the kidneys to ascend. In one in 600 persons the kidneys are fused across the midline. The large U-shaped kidney (horseshoe kidney) lies at the level of the lower lumbar vertebrae, because arteries passing forward from the aorta block its ascent. This condition may be asymptomatic.

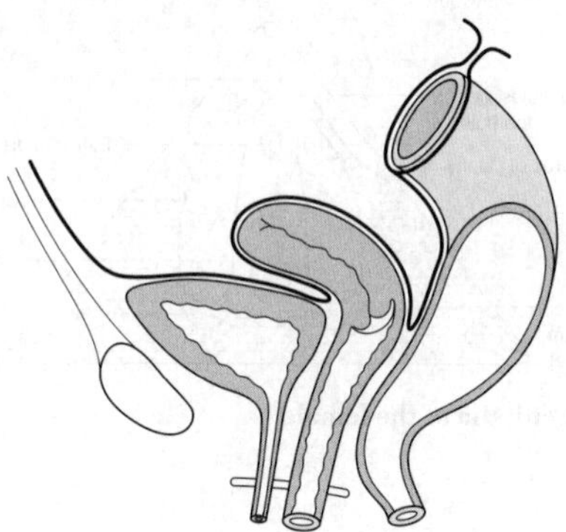

40. Peritoneum in female pelvis in median section.

As the bladder fills it extends above the pubic symphysis—it then lies adjacent to the abdominal wall without the intervention of peritoneum. Consequently the distended bladder may be punctured and catheters inserted without entering the peritoneal cavity: this is a direct method of relieving a blockage of urinary flow due to prostatic enlargement.

Pelvic organ prolapse is a common disorder in women—40 per cent of women over 50 years have some degree of prolapse. Loss of support for the bladder in women from damage to the pelvic floor during childbirth can result in prolapse of the bladder on to the anterior vaginal wall and produce difficulty with passing urine. Similarly, the rectum can prolapse on to the posterior vaginal wall. The treatment is surgical repair of the pelvic floor in severe cases.

The prostate is of considerable medical interest because benign prostatic enlargement is common after middle age, affecting every male who lives long enough. The symptoms are due to obstruction of the urinary outflow, such as poor flow, hesitancy, and an intermittent stream. Obstructions may be relieved surgically by transurethral endoscopic resection (the instrument is passed through the urethra and prostatic tissue is cut away under direct vision).

Ectopic tubal pregnancy may occur when the early developing embryo implants in the uterine tube rather than the uterus. If not diagnosed early, this may result in rupture of the uterine tube and severe bleeding into the abdominopelvic cavity during the first eight weeks of gestation. Such tubal rupture constitutes a threat to the mother's life and is a surgical emergency.

Chapter 7
The limbs

If the foot of man presented no distinguishing peculiarity, the hand, like the corresponding part in other animals would be compelled to share with it the task of carrying the body, and could therefore, not be devoted to the various offices which it is now free to perform. Little right has the hand to say to the foot, 'I have no need of thee'.

G.M. Humphry, Lecturer on anatomy and physiology in the University of Cambridge, 1861

Our upper limbs are free, mobile, and adapted for prehension (the act of grasping or gripping something). Each articulates with the trunk at one small joint, the sternoclavicular joint. Our lower limbs have no such freedom. They have to bear the weight of the body when one walks, runs, jumps, or stands. They are united behind to the vertebral column at the sacroiliac joints and in front to each other at the symphysis pubis.

The upper limb

The upper limb may be divided into the arm (between shoulder and elbow), the forearm (between elbow and wrist), and the hand (below the wrist). Many of the key landmarks in the upper limb can be easily viewed or felt. The clavicle (collar bone) is a subcutaneous bone (lies just under the skin) with a prominent medial end at the sternoclavicular joint and a flat acromial end (at the shoulder). It acts as a strut to force the scapula laterally and backward from the chest wall and is thus vulnerable to fracture if one falls on an outstretched hand or directly on the shoulder. The acromion is the prominent tip of the scapula and if traced back towards the neck it becomes continuous with the spine of the scapula. The greater tubercle of the humerus lies under the cover of deltoid muscle that covers the shoulder joint to produce the roundness of the shoulder.

It is easy to feel the prominence on the inner side of the elbow, the medial epicondyle or 'funny bone'. The ulnar nerve lying behind the medial epicondyle can be felt to slip under the fingers as they are drawn across it, which gives rise to tingling. On the outer side there is the lateral epicondyle, the site or origin of the forearm extensors and for pain in cases of tennis elbow.

The olecranon is the upper end of the ulna. It is the point of the elbow on which one leans. It is protected by a bursa (protective sac of fluid), which may become inflamed (so-called students' elbow). The posterior border of the ulna is subcutaneous—it starts at the olecranon and ends at the wrist as the styloid process of the ulna. The head of the ulna is the rounded medial prominence seen at the wrist on the side of the little finger when the wrist is pronated (palm faces downwards on a table). In contrast the head of the radius is proximal, behind the mobile group of three muscles on the outer side of the elbow. One can feel the radial head moving on rotation of the forearm.

At the wrist, the styloid process of the radius, at the lower end of the bone, lies in the 'anatomical snuffbox', which is visible as a depression on the outer side of the wrist when the thumb is strongly extended. The scaphoid bone can be felt in the snuff box where swelling and tenderness after an injury may indicate a fracture. The radial pulse classically palpated by clinicians can be easily felt at the base of the thumb by pressure against the distal radius. Beyond the wrist, the heads of the metacarpals form the knuckles and are prominent when the fingers are fully flexed. A common punching injury affects the distal part of the fifth metacarpal shaft just below the head (so called boxers' fracture).

Looking at the major muscles, the pectoralis major muscle is the muscular anterior fold of the axilla. It is easily felt if the arm is slightly abducted, or raised, while the posterior fold is formed by teres major and latissimus dorsi, or the swimmer's muscle. Latissimus dorsi brings the outstretched arm from above the head to behind the back. The digitations of serratus anterior (serratus means 'like a saw') can be seen on a muscular person on the chest wall below the axilla. The other end of this muscle is anchored in the scapula and serves to stabilize it against the chest wall. If this function is lost, as occurs if its nerve supply is damaged, the result is 'winging' of the scapula—when the medial border of the bone points backwards on movement, especially when the subject is pushing against resistance. The deltoid muscle forms the curve of the shoulder contour and is a powerful abductor used for lifting the upper limb above the level of the shoulder. The prominent bulge on the anterior aspect of the arm is the biceps brachii muscle. The less prominent bulge on the posterior aspect is due to triceps brachii. The antecubital fossa is the hollow in front of the elbow (Figure 41). When the biceps is relaxed, as by resting the forearm on a table, the pulsations of the brachial artery can be felt adjacent to the biceps tendon and the median nerve, which lies next to the artery and can also be rolled with the finger tips on the underlying brachialis muscle.

At the wrist there are usually two obvious tendons on flexing the joint, flexor carpi radialis and palmaris longus. The palmaris longus is absent in 14 per cent of people on one or both sides. It has a short belly and a long cord-like tendon that attaches to the apex of the palmar aponeurosis, the layer beneath the skin, where the deep fascia is continuous with a sheet of fibrous tissue that divides into four slips, one going to each finger. In its fully

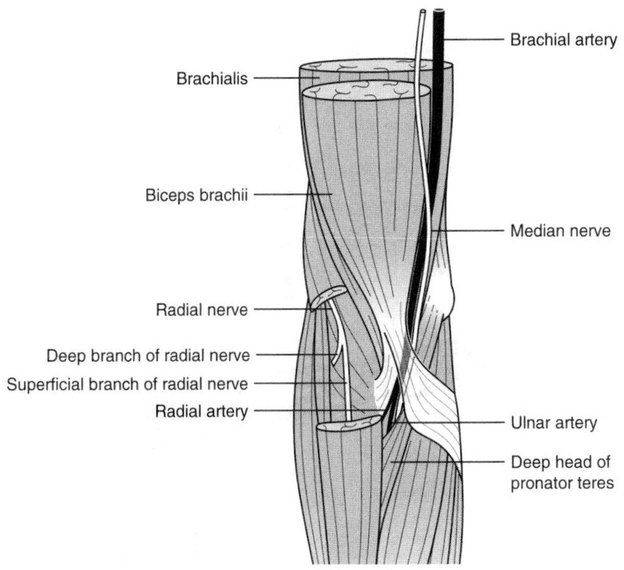

41. The antecubital fossa.

developed functional condition, palmaris longus, in the long distant past, would have flexed the metacarpophalangeal joints, perhaps cupping the hand for drinking. This function in man is now performed by the lumbricals and interossei which are intrinsic muscles of the hand (they arise and insert in the hand) (Figure 42). At the wrist there are two to three flexion creases—the distal crease overlies the proximal row of carpal bones. Beyond this crease is the large lateral thenar eminence (with three intrinsic muscles for the thumb) and the smaller hypothenar eminence (with three similar intrinsic muscles) is in line with the little finger.

The superficial venous drainage of the upper limb is clearly visible in most people with well-used arms. It starts in the venous network on the back of the hand and continues via the basilic vein on the inner side of the forearm with the cephalic vein on the

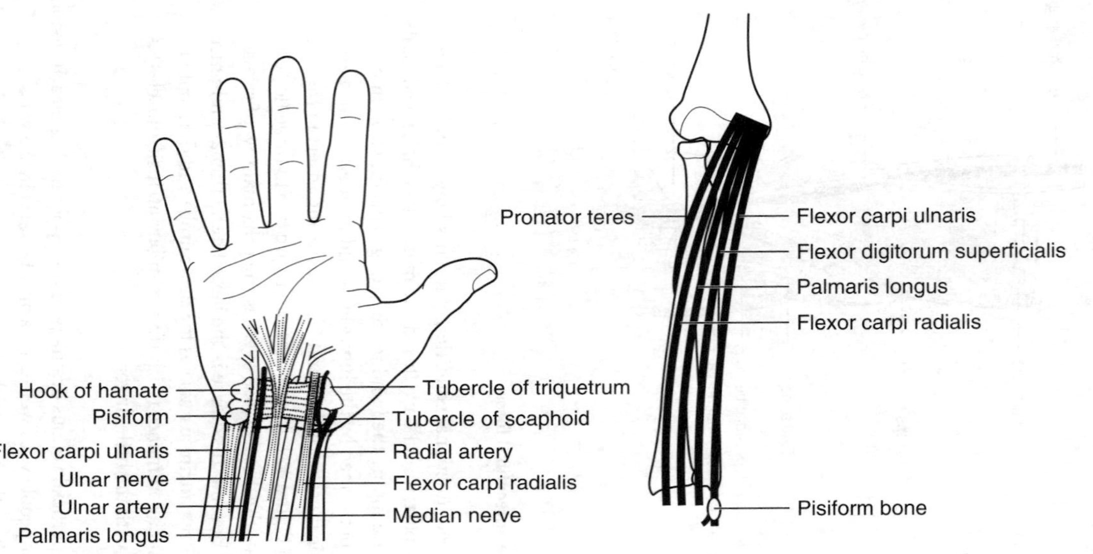

42. Wrist and forearm muscles and tendons.

outer side. These two veins are connected by the median cubital vein at the elbow, which is a common site for collecting blood by venipuncture. The main artery of the upper limb is the subclavian which changes its name as it travels, becoming the axillary in the axilla, and then the brachial artery in the arm—dividing into the radial and ulnar arteries in the cubital fossa. Branches of these two arteries join again the superficial and deep arches of the hand. They form a loop such that the hand's critical blood supply is dual and not restricted. Blood pressure is usually measured at the elbow by temporary occlusion and release of the brachial artery with an inflatable cuff.

There are five main nerves in the upper limb, which develop from the cervical spinal nerves 5–8 and the first thoracic spinal nerve. They form a complex network known as the brachial 'plexus', a confluence of nerves. The median nerve maintains a central position in the limb until it reaches the wrist where it passes through the carpal tunnel, a fibro-osseous tunnel. It is the entrance from forearm to hand at the wrist, formed by a tight anterior fibrous band, the flexor retinaculum, and, posteriorly (at the back), the carpal bones. In the hand, the median nerve supplies the muscles of the bulky thenar eminence, and sensation to the lateral three and a half digits. The ulnar nerve passes around the medial epicondyle (funny bone) and reaches the hand by the little finger for the supply of the majority of intrinsic muscles. A superficial branch continues and supplies sensation to the medial one and a half fingers. The musculocutaneous nerve supplies muscles in the arm and ends as a cutaneous nerve. The radial nerve supplies the muscles on the extensor surfaces of both arm and forearm. In the arm, it winds around the posterior surface of the humerus. It is vulnerable in this situation to stretch lesions if fractures of the humeral shaft occur. The final part of this nerve, the superficial radial nerve (entirely sensory), passes down the forearm with the radial artery to provide sensation for the lateral three and a half fingers (Figure 43).

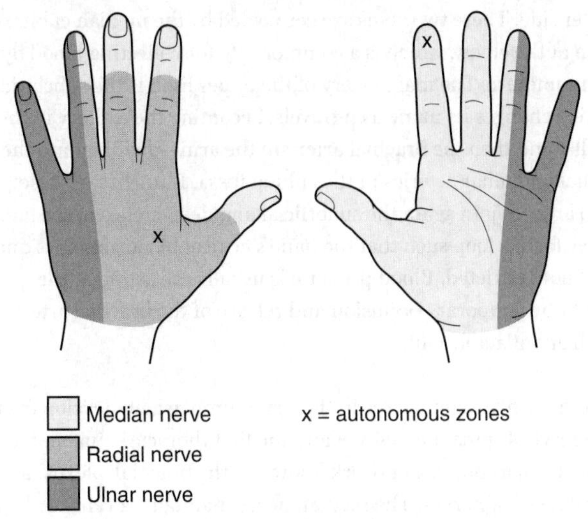

43. Sensory nerve supply of the hand. The autonomous zone is the best site for testing the relevant nerve.

These nerves provide signals to co-ordinate complex muscle movements. In the arm, the brachialis and biceps are the two principal flexors, while all extension of the elbow is produced by the three heads of triceps. The extensor muscles on the back of the forearm arise from the region of the lateral epicondyle of the humerus, while the flexor muscles on the front arise from the region of the medial epicondyle and are stronger than the extensors. There are two pronators and two supinators, which twist the forearm and wrist. It is interesting that Rembrandt, the famous Dutch painter, in the picture called *The Lesson in Anatomy* (1632), mistakenly showed the forearm flexors arising from the lateral side of the arm. The tendons of the hand and finger muscles cross the wrist joint and many of them also pass over the metacarpophalangeal and interphalangeal joints. Both flexors and extensors are arranged in superficial and deep layers. An anatomical arrangement of importance and of beauty is the perforation of the superficial flexor tendons over the middle phalanx of the fingers, so that the deep tendons can insert at the bases of the distal phalanx. Thus the superficial tendons produce flexion at the proximal interphalangeal joints and the deep tendons produce flexion at the distal interphalangeal joints.

The hand is a sense organ capable of transmitting to our brain information concerning the size, weight, texture, and temperature of the objects touched. In order to do this, our upper limbs have developed to satisfy high demands on movement and freedom of action while at the same time being stable enough to give the movements force and precision. The shoulder blade (scapula), clavicle, and shoulder joint form a mechanical unit, but only one joint, the sternoclavicular, connects this complex to the skeleton of the trunk. The clavicle as mentioned is interposed between the thorax and scapula, and articulates with the latter at the acromioclavicular joint. The scapula joins the humerus in the glenohumeral (shoulder) joint. A summation of mobility is therefore established by the three individual joints—all mutually interdependent. The clavicle rotates around the sternum, the scapula around the clavicle, and the humerus around the scapula. In addition, the scapula moves on the chest wall at the 'physiological' scapulothoracic joint. This joint provides one-third of the range of abduction with two-thirds being produced at the shoulder. With all this free motion possible, the main burden of stabilization rests upon the powerful muscles which secure the scapula to the chest wall, such as serratus anterior, latissimus dorsi, and trapezius.

Abduction, adduction, flexion and extension, medial and lateral rotation all occur at the mobile shoulder (glenohumeral joint), which gains mobility at the cost of stability. Anterior dislocation of the shoulder is a common injury. The short muscles on the scapula arranged closely around the humeral head have the important function of retaining the head in its socket. The muscles involved are supraspinatus above, infraspinatus and teres minor behind, and subscapularis in front. Their tendons tend to fuse with the capsule of the joint to produce a 'rotator cuff'. They act as accessory dynamic ligaments. It is interesting that in quadrupeds these muscles insert separately and do not form a rotator cuff. The appearance of the rotator cuff in the evolutionary process may indicate anatomical adaptation to regular overhead activity, such as throwing, and the increased use of the arm away from the sagittal plane (Figure 44).

The elbow is a hinge joint and its main function is to appropriately position the hand in space. This movement is supplemented at

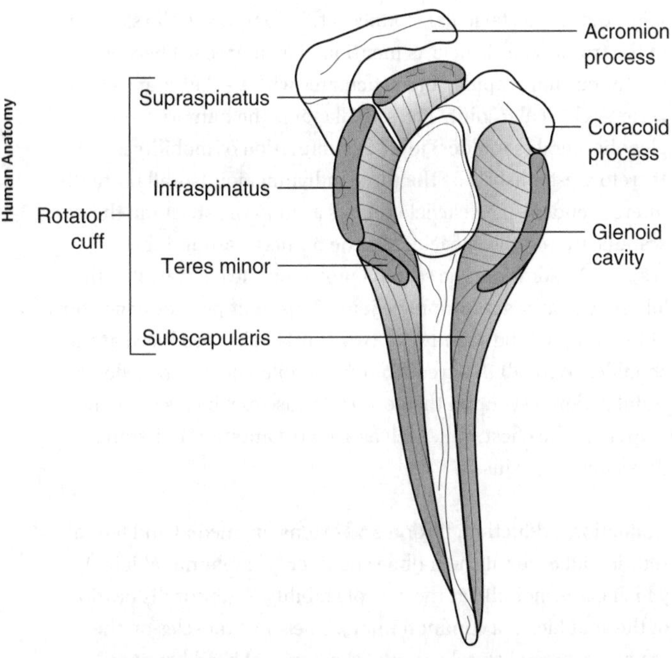

44. View looking into shoulder socket to show the muscles forming the rotator cuff.

the superior and inferior radio-ulnar joints where pronation and supination occur, and thus enable the palm to be placed in the desired position. The wrist joint, which is ellipsoid, allows movements in two planes, flexion and extension and radial and ulnar deviation (side to side movement) and circumduction (all the movements combined).

There are two main patterns of using the hand. These are the precision grip and the power grip. The precision grip takes place between the terminal digital pad of the thumb and the pads of the finger tips. It is employed when delicacy of handling and accuracy of instrumentation are essential and power is a secondary consideration.

The power grip is executed between the surface of the fingers and the palm with the thumb acting as a buttressing and reinforcing agent, as when holding a hammer. Under some conditions the thumb supplies directional control. Precision and power grips are functional concepts, but they are to some extent discrete as far as their nerve supply is concerned. The brunt of paralysis affecting the median nerve falls on the muscles responsible for precision grip, so this is 'the nerve of precision'. The ulnar nerve supplies the bulk of the power grip muscles and can be referred to as the 'nerve of power'. An important adjunct to grasping is extension of the wrist joint. The wrist in 30–40 degrees of extension is in the optimal position for gripping. In full flexion of the wrist the hand can exert only 25 per cent of its power when in full extension. Indeed it has been suggested that the quickest way to disarm an assailant wielding a knife or a gun is to force his wrist into flexion, whereupon he will be forced to drop his weapon.

The lower limb

The lower limb may be divided into the thigh (between hip and knee), the leg (between knee and ankle), and the foot below the ankle. The calf is the term applied to the fleshy posterior part of the leg. During development the lower limbs rotate in the opposite direction to the upper limbs—thus the knees in contrast to the elbows point forward so that the posterior (back) surface of the limb consists of flexor muscles, while extensor muscles occupy the anterior (front) surface.

Looking at the soft tissue landmarks, the prominent buttocks produced by fat over the gluteal region covering gluteus maximus, are more marked in women than in men. Between the bulges is the natal cleft. The buttocks are unique to man and are not found in other primates. Aristotle wrote that man had no tail but he did have buttocks (see role of gluteus maximus in bipedalism in Chapter 8).

On the anterior aspect of the thigh there is the firm prominence of the quadriceps muscle group, key to stability of the knee, underlying the dense fascia lata (the deep fascia of the thigh). On the outer side there is vertical thickening of this fascia, the iliotibial tract which provides insertion for most of gluteus maximus. It is most easily palpable in the lower thigh as a tight band, when the knee is flexed. The superficial inguinal lymph nodes are the only lymph nodes in the body which may normally be palpable. They consist of two sets. There is a proximal or horizontal set lying below and parallel to the inguinal ligament (it separates the abdomen from lower limb), and a distal or vertical set lying alongside the upper end of the great/long saphenous vein. The glands receive drainage not only from the thigh below but from the perineum (area surrounding the outflow tracts of the body) and abdominal wall as well.

The bulge on the back of the thigh is mobile and is thrown up by the hamstring muscles. The diamond shaped hollow behind the knee is the popliteal fossa. It is bounded by the diverging tendons of the hamstring muscles and below by the converging bellies of the gastrocnemius muscle.

The muscles of the leg form a palpably continuous mass extending from the shin, around the lateral and posterior surfaces of the leg, to the medial border of the tibia. The tendons passing from the leg to the foot are readily felt in the ankle region and above the heel. In the latter position the calcaneal tendon (Achilles' tendon) is most obvious. It is the common tendon for four muscle bellies in the calf: the two heads of gastrocnemius (from the Greek, 'belly of the leg'), soleus (whose shape resembles the sole fish), and the vestigial belly of plantaris. The calcaneal tendon is the strongest and thickest tendon in the body and important in human evolution (see Chapter 8). The calf becomes tense when the heel is raised from the ground and thus is more prominent in ladies with high heels, hence the fashion (Figure 45). The plantar muscles do not form clear-cut eminences unlike the thenar and hypothenar muscles of the hand.

Bony landmarks are also present. The symphysis pubis is the lower boundary of the anterior abdominal wall in the midline and lateral to it is the pubic tubercle on either side. The iliac crest on which skirts and trousers rest extends from the anterior superior spine to the posterior superior spine. There are often dimples in the skin over the posterior spines. The iliac tubercle is a prominence of the outer lip of this crest. The ischial tuberosity is best felt on the prone body above the medial part of the fold of the buttock (gluteal sulcus). It is covered by gluteus maximus when one stands. In the sitting position the muscle slips away laterally so that weight is taken directly on the bone. There is fortunately a bursa to protect it and keep us comfortable. The greater trochanter of the femur is felt about 12 centimetres below the middle of the iliac crest and indicates the level of the hip joint. The medial condyle of the femur forms the rounded prominence on the medial side of the knee. The lateral condyle is smaller; between them lies the cavity of the knee joint. The patella, a large sesamoid bone, is in front of the knee. The tuberosity of the tibia is a prominence on the front of the tibia. The leg is connected to the thigh by the strong ligament from

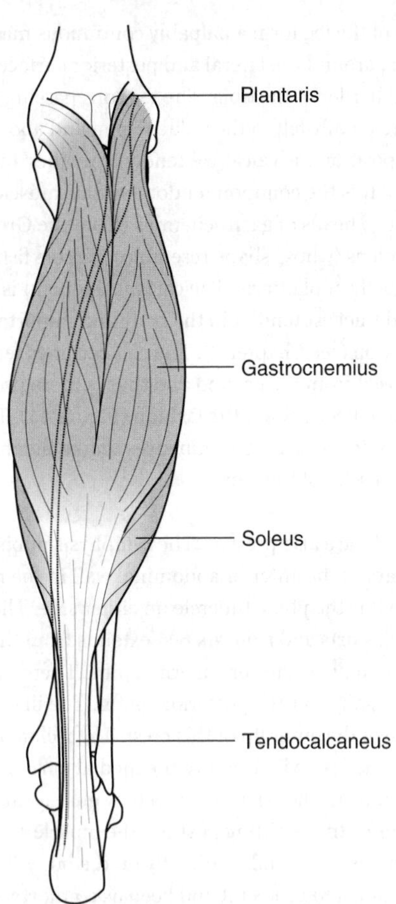

45. The muscles that insert into tendo calcaneus (the Achilles' tendon).

the patella attached to this tuberosity. The surface of the tibia lies just beneath the skin, forming the shin. Below this, the medial prominence of the ankle is the medial malleolus and on the outer side is the lateral malleolus of the fibula, which is the pulley for two tendons passing to the foot.

As in the upper limb, there are two main superficial veins in the lower limb. The great (long) saphenous vein lies on the inner side. It begins at the tibial end of the venous arch on the top of the foot, passes over the front of the medial malleolus, ascends vertically next to the tibia, and continues obliquely up the thigh to enter the femoral vein 2.5 cm below the inguinal ligament, where it pierces the deep fascia. The constant position of the great saphenous vein immediately anterior to the medial malleolus is valuable when emergency access is required as one can perform a 'cut down' (i.e. make a skin incision) on the inner side of the medial malleolus with the certainty of finding a good-calibre vein for transfusion. The great saphenous vein communicates with the deep venous system not only at the groin but also at a number of points along its course through perforating veins. The small (short) saphenous vein is the lateral continuation of the dorsal venous arch of the foot. It passes below and then behind the lateral malleolus, pierces the deep popliteal fascia behind the knee, and ends in the popliteal vein. All the connections of the superficial to deep veins help improve venous return to the heart by muscular contraction within the closed compartments of the limb (a musculovenous pump). There are also eight to ten valves in the long and short saphenous veins to aid this venous return—when these fail, this can lead to the development of 'varicose' veins, skin pigmentation, and ulcers.

Blood supply to the lower limb is through the femoral artery—the distal continuation of the external iliac artery beyond the inguinal ligament. Its first main branch is the profunda femoris artery, which arises below the inguinal ligament and supplies the muscles of the thigh by four perforating arteries which encircle the shaft of the femur so closely that they are usually torn if the shaft is fractured. The femoral artery passes down the thigh to enter the popliteal fossa where it becomes the popliteal artery, finally splitting into the posterior and anterior tibial arteries with companion veins (venae comitantes) (Figure 46). Each artery also has a companion nerve, the tibial and deep fibular nerves,

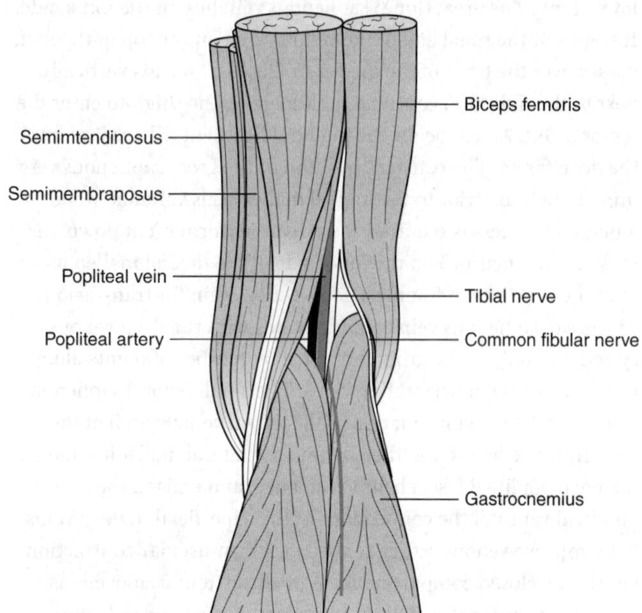

46. The popliteal fossa.

respectively. The posterior tibial artery with the tibial nerve passes through the leg and ends on the inner side of the ankle, the gateway to the foot, where both divide into medial and lateral plantar branches, which supply the foot, the pulse being felt just behind the ankle. The anterior tibial artery continues down to the foot to form the dorsalis pedis artery. This artery, which can also be readily felt and where (normally) a pulse can be located, then runs towards the base of the great toe and passes down into the sole where it joins the lateral plantar artery. Thus these two arteries complete the plantar arch in the sole of the foot with the medial plantar artery, to produce a junction of the two main arteries of the limb just as in the hand. Peripheral vascular disease caused by smoking and hypertension can lead to loss of the pulses in the foot and concomitant restriction of blood supply to the lower limb. The muscles (evertors) attached to the fibula have an arterial supply from the fibular (peroneal) artery, a branch of the posterior tibial artery with a nerve supply from the superficial fibular nerve.

In the thigh the anterior compartment contains flexors of the hip principally (iliopsoas and sartorius (which is known as the tailor's muscle, because it brings the leg into a cross-legged position)) and muscles which extend the knee quadriceps. The medial compartment contains the adductors which bring the leg towards the midline. The posterior compartment comprises the hamstrings which arise from the ischial tuberosity and insert into the leg bones below the knee so that they extend the hip (i.e. straighten it) and flex the knee (i.e. bend it). The lateral compartment includes the glutei (maximus, medius, and minimus, and tensor fasciae latae) whose main overall action is abduction and rotation. The nerves to the lower limb arise from the lumbar and sacral plexuses which supply different compartments of the limb. The femoral nerve supplies the anterior compartment (extending the knee), the obturator nerve supplies the medial (adductor) compartment, the sciatic nerve (the largest nerve in the body) the posterior compartment, and the superior and inferior gluteal nerves supply the lateral compartment.

Unlike the upper limb, the joints of the lower limb must provide stability. The hip is a stable joint of the ball and socket type. The head of the femur forms a true hemisphere of 180 degrees. The socket or acetabulum has the same angular value—thus in midposition the articular surfaces cover each other perfectly. The cartilaginous limbus, which arises from the acetabulum, grasps the head of the femur beyond its equator. This accurate fit is an important factor in the cohesion of the joint. If all the soft tissues were removed, the joint would still be held close together by the atmospheric pressure. Traumatic dislocation of the hip requires considerable violence. Since the ilium, ischium, and pubis each take part in the acetabulum, capsular fibres proceed from each of these to the femur, as the iliofemoral, pubofemoral, and ischiofemoral ligaments. As in the arm there is also a large muscle covering the joint, the gluteus maximus (an extensor), which acts as a lid to the gluteal region covering a number of short lateral rotators behind the hip joint, which function as dynamic stabilizing ligaments, like the rotator cuff of the shoulder.

Gluteus medius and minimus originate from the outer aspect of the ilium and insert into the greater trochanter. They abduct and medially rotate the hip. They are important for walking as they stabilize the pelvis when standing on one leg.

Psoas major arises from the front of the lumbar vertebrae. Iliacus arises on the inside of the iliac bone (the iliac fossa) and inserts into the psoas tendon, which in turn is attached to the lesser trochanter of the femur and functions as the major flexor of the hip joint (Figure 47). The movements permitted at the hip joint are similar to those at the shoulder.

Fractures of the femoral neck are common injuries in the elderly mainly because of osteoporosis. They may be intracapsular (subcapital) or extracapsular (trochanteric). Following such a fracture, the foot lies in a position of marked external rotation and the diagnosis can be made from the end of the trolley in the Accident Department. The blood supply of the head of the femur may be interrupted in intracapsular fractures leading to bone death (avascular necrosis). Osteoarthritis of the hip is nowadays a common disorder and a clear disadvantage of our upright posture, but fortunately it is amenable to total hip replacement.

The main movements at the knee are flexion and extension, but as the joint is, in contrast to the elbow, a modified hinge joint, some degree of axial rotation is also permitted when the knee is in the position of flexion and semiflexion. The collateral ligaments

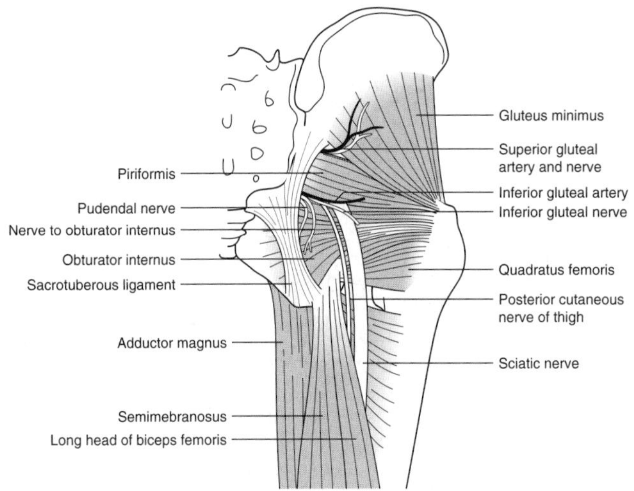

47. Gluteal region and hamstring origins.

(i.e. on the sides of the joint) are attached above the epicondyles, which project like hubs on wheels from the femoral condyles. When the knee is flexed, the collateral ligaments are slack and allow medial and lateral rotation. The medial collateral ligament has a superficial and deep part. The deep part is triangular in shape and is attached to the margins of the tibial condyle and medial meniscus in the joint.

In order to improve the fit of the femoral and tibial surfaces there are two menisci (semilunar fibrocartilages) that rest on the tibial condyles. The menisci are bound to the margins of the condyles by ligaments and attached to the nonarticular part of the upper surface of the tibia. The cruciate ligaments cross in the centre of the knee and maintain forward stability. They take their names from their tibial attachments and cross each other obliquely like the limbs of a St Andrew's cross (Figure 48). Injuries of the knee are common and occur during rotation in flexion, as is used in football. This produces a variety of injuries depending on the

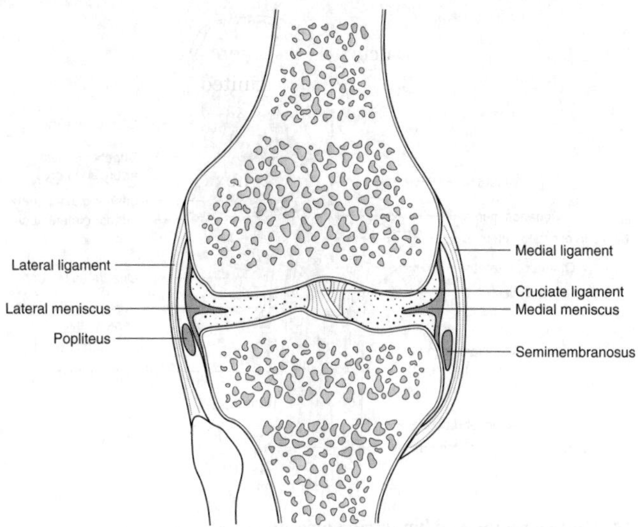

48. Coronal section of knee joint.

forces involved. The anterior cruciate is the most common ligamentous injury. The menisci may sustain longitudinal splits (bucket-handle tears). Because of its attachment to the medial ligament the medial meniscus is more commonly affected than the lateral one.

The ankle is a hinge joint, allowing dorsiflexion (extension) and plantar flexion (flexion). The talus sits between the tibia and fibula. They are joined together by a strong interosseous ligament reinforced by anterior and posterior distal tibiofibular ligaments, which allow some suppleness of the joint as it widens and the broad end of the talus engages in dorsiflexion at 90 degrees, its most stable position. There is a strong medial collateral ligament (deltoid ligament). The lateral collateral ligament, however, is in three parts and vulnerable to injury. The anterior talofibular ligament is a band that passes from the anterior border of the lateral malleolus to the neck of the talus. It is the weakest ligament of the joint and often torn in ankle sprains. This is significant as it has been calculated that there is one sprain per day per 10,000 of the population in the United Kingdom. The calcaneofibular ligament is a cord that resembles the lateral ligament of the knee, and when torn with the anterior ligament will result in recurrent ankle instability. The posterior talofibular ligament passes medially and backwards, and is not usually torn.

Plantarflexion is produced by gastrocnemius and soleus via the calcaneal tendon.

The two muscles have different functions. The soleus acts predominantly as a postural muscle to prevent the body falling forward when standing at rest. In contrast the gastrocnemius produces vigorous propulsive movements for running and jumping. As the calcaneal tendon transmits forces that are up to seven times the body weight during running, it is not surprising that it commonly ruptures in middle-aged athletes as in the fathers' race on school sports day. Dorsiflexion is mainly due to the pull of the long extensors of the great and lesser toes. Our feet are dynamic springs that function best when we are active. In contrast protracted standing such as occurs when visiting an art gallery may be very uncomfortable. The joints of the foot ensure that the foot is in the optimal position for weight bearing. This is of particular importance on uneven ground. The talus plays a unique role in this as it links the leg and foot. The ankle joint is its proximal connection to the leg and the subtalar (sometimes called the lower ankle joint) its distal connection to the foot.

Inversion moves the sole of the foot to face medially and eversion turns the sole laterally. These movements occur at the subtalar and midtarsal joints. Clinicians and runners often use the terms pronation and supination in relation to the foot. These are not synonymous with inversion and eversion. Rather, they apply to the whole sole of the foot, which is usually weight bearing. The flat foot with no obvious arch is described as pronated. Pronation is a

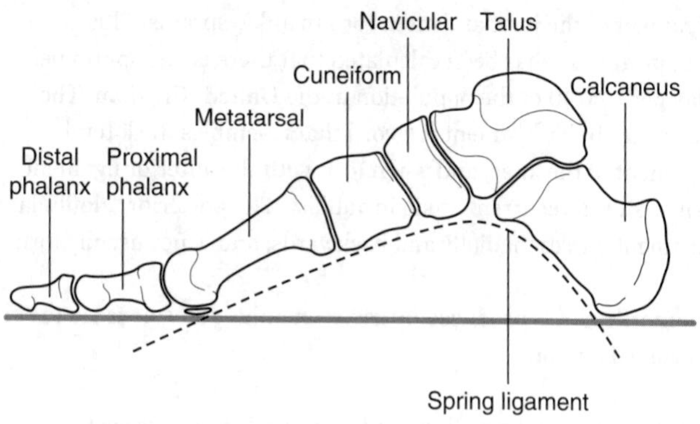

49. The medial longitudinal arch.

combination of dorsiflexion, abduction, and eversion at the various joints around the foot.

With regard to the arches of the foot although a transverse arch is traditionally described, it is not functional but rather part of the architecture. Similarly the lateral longitudinal arch is low and contributes little to the spring action of the foot. The key medial longitudinal arch is formed by the calcaneus, talus, navicular, three cuneiforms, and three medial metatarsals. At its summit (which is the junction of its posterior one-third with its anterior two-thirds) lies the head of the talus (Figure 49). The plantar aponeurosis or central part of the plantar fascia in lower mammals was continuous with the plantaris muscle. In man the muscle is vestigial and the aponeurosis has been modified to act as a strong tie for the longitudinal arches of the foot. At the rear, it is attached to the calcaneus and anteriorly it splits into five bands one for each digit. It is a key factor in the function of the medial longitudinal arch. Dorsiflexion of the toes produces tightening of the plantar aponeurosis which lengthens by about 9–12 per cent. This has been described as the 'windlass action' (Figure 50). It causes the arch to rise, the heel to invert, and the tibia to rotate. This process is passive without muscle action. The strain energy in the raised arch

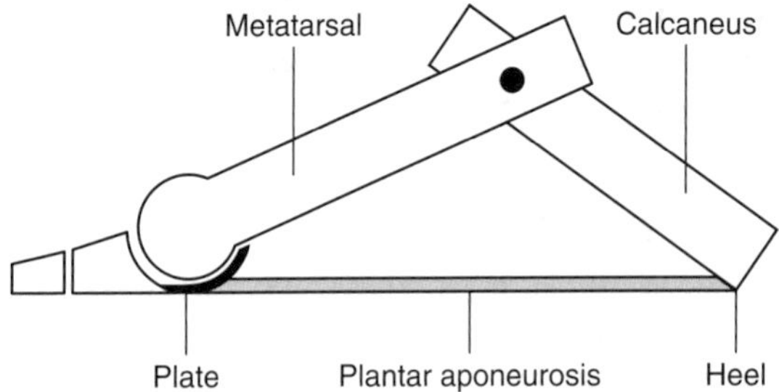

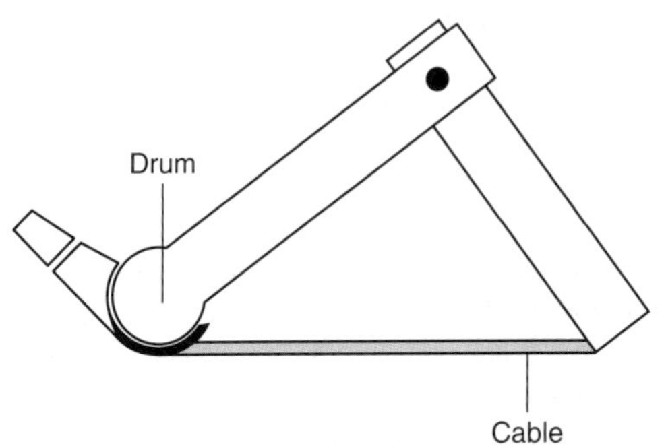

50. The windlass mechanism of plantar aponeurosis.

produces the spring in the foot when the next step is started, as the foot turns inwards and becomes a strong lever with a stable midtarsal joint for 'toe off'. The support of the arch by the plantar aponeurosis is supplemented by deep ligaments on the sole, the long and short plantar ligaments, and the spring ligament. It has been demonstrated that muscles play no important role in the normal static support of the longitudinal arches. With regard to the diagnosis of 'flat feet' the height of the medial longitudinal arch is not of significance. This is exemplified by the Olympic sprinter Usain Bolt who has no obvious longitudinal arch when standing still. Nevertheless his windlass action works normally when he is walking or running. Tightening the windlass by passive dorsiflexion of the great toe, or standing on tip toe will reveal whether or not the windlass is working. Mobile flat feet are normal and have been grossly overdiagnosed as pathological in the past.

When things go wrong

Carpal tunnel syndrome is a painful condition when the median nerve is compressed in the fibro-osseous carpal tunnel at the wrist, because of pressure due to increase of contents of the tunnel. This may occur in pregnancy due to the increased fluid content of the body or be due to thickenening of the tendon sheaths of the tendons passing through the tunnel.

Minor developmental defects are relatively common, but major limb malformations are rare. Lobster cleft hand or foot is the absence of one or more central digital rays or digits. Thus the hand or foot is divided into two parts that oppose each other like lobster claws. In club hand or congenital absence of part or the whole of the radius, the hand deviates radially and the ulna bows with the concavity on the radial side. Polydactyly or supernumerary fingers or toes are common. Similarly fusion of the fingers or toes is one of the most common limb deformations, and is more frequent in the foot than in the hand (syndactyly). Syndactyly occurs because of failure of cell death (or apoptosis) of the tissue between the developing digits.

Clubfoot is relatively common and occurs about twice as frequently in boys. At birth the sole of the foot is turned medially and the foot is adducted and plantarflexed. If untreated the child will walk on the outer borders of the feet.

Congenital dislocation of the hip is now formally known as developmental dysplasia of the hip. Girls are more commonly affected than boys, the ratio being about 7:1. The capsule of the hip is abnormally lax at birth and there is underdevelopment of the acetabulum and head of the femur. The actual dislocation almost always occurs after birth.

Chapter 8
Man, the tottering biped

Nothing in biology makes sense except in the light of evolution.

Theodore Dobzhansky

Bipedalism is no easy accomplishment. It requires a fundamental reconstruction of our anatomy, particularly the foot and pelvis.

Stephen Jay Gould

We have looked at the complex anatomy of humans in the previous chapters. In order to understand how we reached our present state it is necessary to explore how the adoption of a two-legged gait led to widespread and profound changes in our anatomy.

Africa sprang its greatest surprise on the world of anthropology in 1925 when Raymond Dart, newly appointed Professor of Anatomy at the medical school in Johannesburg, published the first account of the skull of a fossil child. The find was a geographical jolt for the world, which was just becoming used to the idea of Asia being the home of mankind, when out of the blue the Taung child arrived to stake out Africa's claim. The base of the cranium showed evidence that the head had been rather well-balanced on what must have been a virtually upright spine, in contrast with the position in apes in which the head hangs forward from an obliquely postured spine. Dart believed that this skull discovered in Taung in the northern Cape did not belong to the family of apes, and he assigned the specimen to a new genus and species *Australopithecus africanus* (southern ape of Africa)—the link between apes and man. Africa was proposed as the cradle of humanity. With further finds, especially in the eastern part of the continent, Africa has since remained the centre of research for human origins.

Many lines of evidence suggest that climate change spurred selection for bipedalism in order to improve early hominins' ability to acquire food in times of scarcity. Bipedal hominins probably would not have come about if knuckle walking, fruit-eating apes had not previously evolved to live in the African rainforest. Perhaps they were forced to become bipeds as the central African forests diminished in size. Woodlands and forests became patchier and were separated by more savannas and grassland. There was a need to forage in open woodland with widely scattered food resources. Bipedal walking and running was an energetically efficient way for a large primate to travel long distances over flat country. Darwin speculated in 1871 that of all the characteristics that make humans distinct, it was bipedalism rather than big brains, language, or use of tools that first set the human lineage off on its separate path from the other apes. Darwin reasoned that it was bipedalism that initially emancipated the hands from locomotion. This allowed natural selection to favour additional capabilities such as making and using tools. In turn these capabilities selected for bigger brains, language, and other cognitive skills.

Hominin axial skeletons show many derived adaptations for bipedalism including an elongated lumbar region both in regard to the number of vertebrae (apes usually have three or four and humans five), as well as a marked posterior concavity (wedged lumbar vertebrae curve the lower lumbar spine inward). The benefits of walking and standing upright must have outweighed the costs at every evolutionary stage. Almost all discussions of hominin bipedal locomotion make the implicit assumption that they were adapted for bipedal walking. However, recently it has been suggested that the selection may not have been for walking, but for bipedal endurance running. Humans, of all our planet's species, are the best hot weather distance runners. Several of the morphological features of the modern human foot are more easily explained as adaptations to running rather than to walking. The first direct evidence of bipedalism comes from the hominin footprint trails at Laetoli, Tanzania. The footprints were made in layers of volcanic ash called tuff, dated 3.8–3.4 million years ago. The number of individuals responsible for the trails is contested. Some researchers interpret them as the footprints of three individuals, two walking side by side followed by a third hominin directly behind one of the individuals. Other workers interpret the trails as the footprints of two individuals walking side by side. Most researchers agree that the Laetoli footprints are evidence of a hominin (*Australopithecus afarensis*) with a fully adducted great toe as in modern man rather than the abducted prehensile toe of the ape, in other words instead of a great toe similar to the thumb, which is separated from the index finger with a wide gap, useful for clinging to branches, the great toe was positioned parallel with the second toe. In addition several workers suggest that the Laetoli footprints were made by a hominin with a medial longitudinal arch. This evidence shows a foot that resembles the feet of modern humans. By this time it is likely that most of the distinctive elements of our feet were present. These include a short compact talus, modest but definite longitudinal and transverse arches, and midfoot joints that 'lock' to convert the tarsus into a stiff lever that can transmit the force generated by the calf muscles to produce a powerful 'toe off'. The barefoot Australopithecine as seen in 'Lucy' (*Australopithecus afarensis*) was capable of walking, but the skeletal design of this type of foot changed at the time of the Pliocene/Pleistocene boundary about two million years ago. At this stage, longer hind limbs and shorter toes developed, and if the endurance hypothesis is correct the evolution of these features along with a fully arched foot is linked directly to barefoot running as an integral part of hunting.

Bipedal gaits are inherently unsteady, hence the description 'tottering biped' by Edward Hooton, a famous anthropologist at Harvard University. Several differences between walking and running call for special mechanisms to help ensure stabilization and balance. Humans have a number of derived features that enhance stabilization of the trunk. There are expanded areas on the sacrum and posterior iliac spine for the attachment of the large erector spinae muscles, and a greatly enlarged gluteus maximus (Figure 51). The latter muscle, whose increased size is among the most distinctive of all human features, is strongly recruited in running at all speeds, but not in walking on level surfaces. In addition the transverse processes of the sacrum are relatively larger in man than in the Australopithecus hominin, suggesting a more stable sacroiliac joint.

Running also poses problems for the stabilization of the head. Unlike quadrupeds, humans have vertically oriented necks that are less able to counteract the greater tendency of the head to pitch forward during running than walking. The radius of the posterior semicircular canal, part of the balance system located in the ear, is significantly larger in man than in chimpanzees and Australopithecines. This presumably increases sensory perception to head pitching in the sagittal plane. Another structural modification here relevant to running is the ligamentum nuchae (nuchal ligament,) running down the back of the neck, a feature of animals that are either cursorial (i.e. adapted specifically for running) such as dogs, horses, and hares; or have massive heads such as elephants. A nuchal ligament is absent in chimpanzees and Australopithecines. This ligament is really a tendon-like structure that evolved independently in humans and the animals mentioned above. In humans the ligament has two parts. The more superficial is a thickened seam or raphe, sometimes the

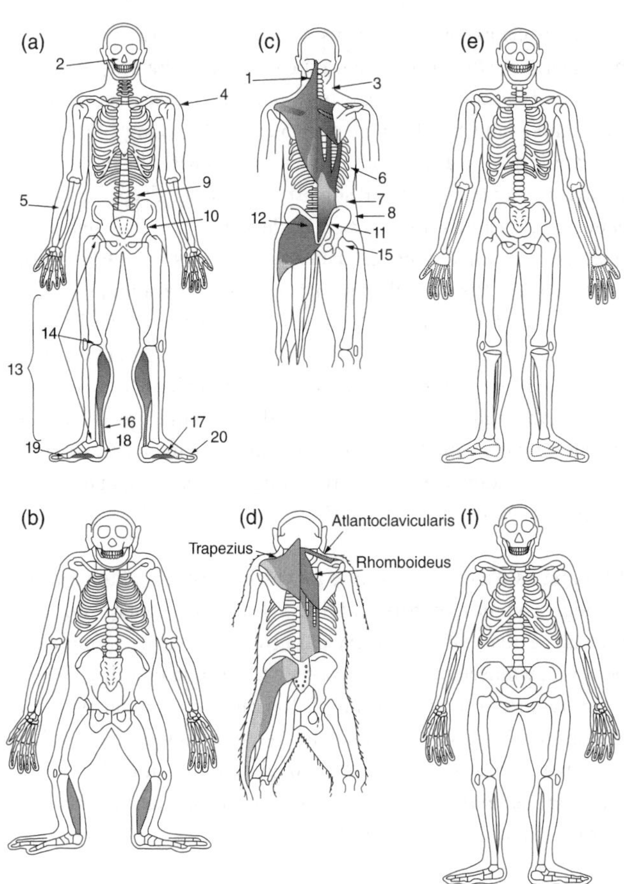

51. Diagram to show changes that occurred with the development of bipedalism. (a), (c) Anterior and posterior views of human with arched foot, calcaneal tendon and large gluteus maximus; (b), (d) Anterior and posterior views of chimpanzee. Labelled muscles connect head and neck to pectoral girdle and are reduced or absent in humans; (e) Reconstruction of Homo erectus; (f) Reconstruction of Australopithecus afarensis.

diameter of a pencil, which runs between the external occipital protuberance (the projection on the most posterior part of the skull) and the spinous process of the seventh cervical vertebra. This tendon-like band, which does not insert into any of the intervening cervical vertebrae, interdigitates with the upper cranial part of trapezius and sometimes with the deeper muscle, semispinalis capitis. In contrast the deeper part of the nuchal ligament is a fascial septum that extends from the midline of the occiput below the nuchal crest on the skull to the spinous processes and interspinous ligaments of the seven cervical vertebrae. Together the two parts of the nuchal ligament create a partially elastic connection in the midsagittal plane between the occiput, the upper neck extensors, and the shoulder girdle. Several studies on quadrupeds have demonstrated the nuchal ligament acts like a tendon, capable of stretching and recoiling in a spring-like manner.

Running uses a compliant limb in which muscles and tendons in the legs sequentially store and then release strain energy during the stance phase of the gait cycle. In contrast to apes, human legs have many muscles that can generate force economically. The most important of these springs is the calcaneal tendon, which is absent in primates; other elongated tendons include the iliotibial tract, which receives the insertion of gluteus maximus. The medial longitudinal arch of the foot with the elastic plantar aponeurosis provides another important spring.

Early man's capacity to accomplish successful hunting was based on his ability to run long distances and his resistance to heat. We are the only one of the species of ape and monkey without body hair, and in addition we have two million sweat glands. No mammal sweats as much as humans. Bipedalism reduces the surface area exposed to direct solar radiation; and increased vertical height maximizes the total skin surface for evaporation and convection. Animals such as wildebeest cannot cope with the increased heat generated by running in a hot climate. They become exhausted and die from hyperthermia. Most medium to large animals rely on evaporative cooling to maintain body temperature while running. This is accomplished by two separate mechanisms, respiratory evaporation occurring at the nasal mucosa, mouth and tongue surfaces (panting), and evaporation of sweat from the general body surface, which is only possible in man, horses, and camels. The other animals cannot gallop and pant. Because of his endurance capacity, primitive man could continue running after animals until they succumbed to heat exhaustion, when it was relatively easy to kill them with primitive weapons.

The evolution of childbirth

The shape of the modern human pelvis represents a compromise between its obstetric obligations and the requirements of support, and walking and running. The fossil record presents an interesting glimpse into the process by which this was achieved. Two reasonably complete, though distorted, female pelves have been discovered and described from Australopithecines. The first is assigned to *Australopithecus africanus* and the second to the famous 'Lucy' skeleton of *Australopithecus afarensis*. Both define a birth canal that is very wide from side to side, but narrow from front to back. This pattern has been interpreted as having been driven by locomotor rather than obstetric demands. An Australopthecine baby may have had to enter and leave the birth canal sideways. Bipedalism requires shifting the position of the sacroiliac joint as close to the acetabulum as possible. Widening the pelvis laterally gives more support to the abductors of the hip. It is difficult to know at what point the fit between birth canal and foetal size became so tight as to require foetal rotations during delivery. With the appearance of *Homo* the brain increased substantially and the pelvis began to assume its modern shape. Several partial rotations were then required to enable the head and then the shoulders to negotiate the constrictions.

The upper limb

Five digits is the ancestral trait for mammals (pentadactyl). The typical primate hand is characterized by a diminutive thumb in combination with long curved fingers. In both monkeys and apes, the thumb is generally reduced in size and unable to contact the tips of the other fingers. The main flexor muscle of the thumb, flexor pollicis longus, is rudimentary in the ape. In contrast the human hand has a much more muscular, mobile, and fully opposable thumb, which is combined with straight fingers. Where the human hand is used solely for manipulation, the ape hand is used both for manipulation and walking. The African ape, the chimpanzee, and the gorilla use their hands in a knuckle walking posture when moving on the ground. The interphalangeal joints of the fingers are flexed and the weight is born on the heads of the middle phalanges.

Fine manipulative skills and a dependence on tools are hallmarks of the human species. The degree of mobility of the human hand is far greater than in other primates. The motor and sensory portions of our brains greatly expanded for perception and control of our hands, and this is reflected in a high density of nerve endings in the muscles joints and skin of the hand. The earliest evidence for human manipulative abilities comes from hand and arm bones of Australopithecines, especially the East African species *Australopithecus afarensis*. The first major changes appear with the emergence of the genus *Homo*, in the species *Homo habilis*. Fossilized hand bones indicate a modern human pattern of joint mobility of the thumb. *Homo habilis* shows for the first time major expansion and probable reorganization of the human brain, fully terrestrial bipedalism, and the manufacture of tools. With the emergence and spread of anatomically modern humans, *Homo sapiens*, 100,000 years ago came the appearance of manipulative skills similar to our own. The hand and arm of these early modern people are indistinguishable from those of athletic living people in terms of articular configurations, proportions of the hand bones, and of the upper limbs and levels of muscularity.

Human manipulative prowess has within the past 40,000 years become one of the dominant aspects of our biological and cultural adaptation.

Evolution has not stopped but continues today as it always has. Species arise and species become extinct. According to Professor Chris Stringer, Head of the Human Origins Group, at the Natural History Museum in London:

> The future of our species is as unpredictable as was our past. Mammalian species generally have lifetimes measured in the hundreds of thousands or the low millions of years or so. On that basis, we could expect a long but finite future. Our chances of survival seem to rest partly in our own hands.

Further reading

Gray's Anatomy edited by Susan Standring. Churchill Livingstone 39th edition 2005.

Grant's Method of Anatomy by John Charles Boileau Grant, John V. Basmajian, and Charles E. Slonecker. Williams and Wilkins 1989.

Clinical Anatomy by Harold Ellis and Vishy Mahadewan. Blackwell Publishing 13th edition 2013.

The Story of the Human Body by Daniel Lieberman. Allen Lane 2013.

Hands by John Napier. Revised by Russell H. Tuttle. Princeton University Press 1980.

Mutants: On the Form, Varieties and Errors of the Human Body. Arman Marie Leroi, Harper Perennial 2005.

A Short History of Anatomy and Physiology by Charles Singer. Dover Publications 1957.

The Tissues of the Body by W. E. Le Gros Clark. Oxford University Press 6th edition 1975.

The Human Strategy: An Evolutionary Perspective on Human Anatomy by John H. Langdon. Oxford University Press 2005.

The Human Foot: A Companion to Clinical Studies by Leslie Klenerman and Bernard Wood. Springer 2006.

An Introduction to Human Evolutionary Anatomy by Leslie Aiello and Christopher Dean. Academic Press 2002.

Vertebrates: Comparative Anatomy, Function, Evolution by Kenneth V. Kardong. McGraw-Hill International 5th edition 2009.

African Exodus by Chris Stringer and Robin McKie. Jonathan Cape 1996.

Index

C

D

E

J

K

L

M

N

O

P

R

S

T

U

V

W

ONLINE CATALOGUE

A Very Short Introduction

Our online catalogue is designed to make it easy to find your ideal Very Short Introduction. View the entire collection by subject area, watch author videos, read sample chapters, and download reading guides.

http://fds.oup.com/www.oup.co.uk/general/vsi/index.html

SOCIAL MEDIA
Very Short Introduction

Join our community

www.oup.com/vsi

- Join us online at the official Very Short Introductions **Facebook** page.
- Access the thoughts and musings of our authors with our online **blog**.
- Sign up for our monthly **e-newsletter** to receive information on all new titles publishing that month.
- Browse the full range of Very Short Introductions online.
- Read **extracts** from the Introductions for free.
- Visit our library of **Reading Guides**. These guides, written by our expert authors will help you to question again, why you think what you think.
- If you are a teacher or lecturer you can order inspection copies quickly and simply via our website.

SLEEP

A Very Short Introduction

Russell G. Foster & Steven W. Lockley

Why do we need sleep? What happens when we don't get enough? From the biology and psychology of sleep and the history of sleep in science, art, and literature; to the impact of a 24/7 society and the role of society in causing sleep disruption, this *Very Short Introduction* addresses the biological and psychological aspects of sleep, providing a basic understanding of what sleep is and how it is measured, looking at sleep through the human lifespan and the causes and consequences of major sleep disorders. Russell G. Foster and Steven W. Lockley go on to consider the impact of modern society, examining the relationship between sleep and work hours, and the impact of our modern lifestyle.

www.oup.com/vsi